D1760134

624·078·2 K91

0548813-3

WITHDRAWN
FROM STOCK

b
B
a
Re
54

...books may be renewed b...

DUE FOR RETURN

20 MAR 199...

- 2 DEC 1999

DUE FOR RETURN

1 9 OCT 19...

OCT 2000

Guide to Design Criteria for Bolted and Riveted Joints

Guide to Design Criteria for Bolted and Riveted Joints

Second Edition

Geoffrey L. Kulak

John W. Fisher

John H. A. Struik

A WILEY-INTERSCIENCE PUBLICATION

JOHN WILEY & SONS

New York · Chichester · Brisbane · Toronto · Singapore

Copyright © 1987 by John Wiley & Sons, Inc.

All rights reserved. Published simultaneously in Canada.

Reproduction or translation of any part of this work
beyond that permitted by Section 107 or 108 of the
1976 United States Copyright Act without the permission
of the copyright owner is unlawful. Requests for
permission or further information should be addressed to
the Permissions Department, John Wiley & Sons, Inc.

Library of Congress Cataloging in Publication Data:

Fisher, John W., 1931-
  Guide to design criteria for bolted and riveted joints.

  ''A Wiley-Interscience publication.''
  Bibliography: p.
  Includes index.
  1. Bolted joints. 2. Riveted joints. I. Kulak,
Geoffrey. II. Struik, John H. A., 1942–
III. Title.

TA492.B63F56 1987   671.5   86-22390
ISBN 0-471-83791-1

Printed in the United States of America

10 9 8 7 6 5 4 3 2 1

Foreword

Since the first edition of this book was published in 1974, numerous international studies on the strength and performance of bolted connections have been conducted. In the same period, the Research Council on Structural Connections has developed two new specifications for structural joints using ASTM A325 or A490 bolts, one based on allowable stress principles and the other on a load factor and resistance design philosophy. In addition, the Research Council has approved and published "A Test Method to Determine the Slip Coefficient for Coatings used in Bolted Joints." This second, updated, edition has been prepared and approved by the Research Council.

Formed in 1947 and formerly known as the Research Council on Riveted and Bolted Structural Connections, the Research Council on Structural Connections is a nonprofit technical body composed of consulting engineers, educational institutions, trade associations and government agencies, and individual members. Its membership is classified according to interest either as producers, users, or general interest, and its purpose is to promote technical information and the knowledge of economical, efficient and safe practices related to the design and installation of bolted structural connections.

The Research Council's Specifications have been endorsed by the Industrial Fasteners Institute and endorsed and published by the American Institute of Steel Construction, and they form the basis of the technical requirements of other national standards.

Research Council on
Structural Connections

Preface

The impetus for the preparation of a second edition of the *Guide to Design Criteria for Bolted and Riveted Joints* has been the enthusiastic reception of the original version and the continued citation for over a decade of that book as a source of information regarding the design of bolted connections. There certainly has been no other single reference in this area that has been so accepted by designers, teachers, students, and specification writers.

Of course, a great deal of research into the behavior of bolted connections has been conducted since the publication of the original *Guide*. Indeed, that publication itself identified areas of concern and areas in which information was lacking, and it thereby stimulated research. The success of the original *Guide* and the amount of significant research that has been done since its publication indicated that a revised and updated version was desirable. The authors are grateful to the Research Council on Structural Connections for supporting the preparation of a second edition and for underwriting the costs involved. Many council members offered suggestions, and the efforts of the council were coordinated by M. I. Gilmor of the Canadian Institute of Steel Construction. Geoffrey Kulak also particularly wishes to express thanks to Professor J.-C. Badoux, Director of ICOM (Institute de la Construction Métallique) at École Polytechnique Fédérale de Lausanne in Switzerland. The majority of the revisions contained in the second edition were prepared while that author was a visiting professor at the Institute.

Readers of the original *Guide* will find that the second edition follows the same framework. After introductory chapters treating the historical background of high-strength bolts and rivets, the behavior of individual fasteners is introduced. This is followed by descriptions of the behavior of fasteners in the various types of connections that are encountered in structural engineering practice. Throughout the book, new data have been used to update the information originally presented or to present thoughts in areas that were not covered at all in the earlier edition. There has been a great deal of new work for the authors to draw on. For example, there have been significant improvements in our knowledge of the behavior of slip-resistance connections, fatigue of bolted and riveted connections, beam-to-column connections, and so on. The authors are grateful to researchers throughout the world who have shared their results so willingly.

GEOFFREY L. KULAK
JOHN W. FISHER
JOHN H. A. STRUIK

Edmonton, Alberta, Canada
Bethlehem, Pennsylvania
January 1987

Preface to First Edition

This book provides a state-of-the-art summary of the experimental and theoretical studies undertaken to provide an understanding of the behavior and strength of riveted and bolted structural joints. Design criteria have been developed on the basis of this information and should be beneficial to designers, teachers, students, and specification-writing bodies.

The book is intended to provide a comprehensive source of information on bolted and riveted structural joints as well as an explanation of their behavior under various load conditions. Design recommendations are provided for both allowable stress design and load factor design. In both cases, major consideration is given to the fundamental behavior of the joint and its ultimate capacity.

The work on this manuscript was carried out at Fritz Engineering Laboratory, Lehigh University, Bethlehem, Pa. The Research Council on Riveted and Bolted Structural Joints sponsored the project from its inception in 1969.

The work has been guided by the Councils Committee on Specifications under the chairmanship of Dr. Theodore R. Higgins. Other members of the committee include: R. S. Belford, E. Chesson, Jr., M. F. Godfrey, F. E. Graves, R. M. Harris, H. A. Krentz, F. R. Ling, W. H. Munse, W. Pressler, E. J. Ruble, J. L. Rumpf, T. W. Spilman, F. Stahl, and W. M. Thatcher. The authors are grateful for the advice and guidance provided by the committee. Many helpful suggestions were made during the preparation of the manuscript. Sincere appreciation is also due the Research Council on Riveted and Bolted Structural Joints and Lehigh University for supporting this work.

A book of this magnitude would not have been possible without the assistance of the many organizations who have sponsored research on riveted and bolted structural joints at Fritz Engineering Laboratory. Much of the research on the behavior of riveted and bolted structural joints that was conducted at Fritz Engineering Laboratory provided background for this study and was drawn on extensively. Those sponsoring this work include the American Institute of Steel Construction, the Pennsylvania Department of Transportation, the Research Council on Riveted and Bolted Structural Joints, the United States Department of Transportation-Federal Highway Administration, and the Louisiana Department of Transportation.

The authors are particularly grateful for the advice provided by Dr. Theodore R. Higgins and Dr. Geoffrey L. Kulak. Many helpful suggestions were provided that greatly improved the manuscript and design recommendations.

The manuscript was typed by Mrs. Charlotte Yost, and her assistance with the many phases of the preparation of the manuscript is appreciated. Acknowledgment is also due Mary Ann Yost for her assistance with the preparation of the various indexes provided in this book and other resource material. Many organizations have given permission to reproduce graphs, tables, and photographs. This permission is appreciated and credit is given at the appropriate place.

JOHN W. FISHER
JOHN H. A. STRUIK

Bethlehem, Pennsylvania
July 1973

Contents

Chapter One
Introduction

1.1 PURPOSE AND SCOPE

The purpose of this book is to provide background information and criteria that can be used as a guide to the improvement of existing design procedures and specifications for bolted and riveted joints. To achieve this goal, extensive research work performed in the United States, Canada, Australia, Germany, the Netherlands, England, Norway, Japan, and elsewhere was reviewed.

Among the criteria considered as a basis for design was an evaluation of the load versus deformation characteristics of the component parts of the joint. The major emphasis was placed on the behavior of structural joints connected by ASTM A325 or A490 high-strength bolts. The joint materials considered ranged from structural carbon steel with a specified yield stress between 33 and 36 ksi (227 and 248 MPa) to quenched and tempered alloy steel with a yield stress ranging from 90 to 100 ksi (620 to 689 MPa).

The different types of fasteners, connections, loading conditions, and design procedures are discussed briefly in the first two chapters. Chapters 3 and 4 deal with the behavior of individual fasteners under various loading conditions. Chapter 5 describes the behavior, analysis, and design of symmetric butt splices. Special types of joints such as truss-type connections, shingle joints, beam or girder splices, and beam-to-column connections are discussed in subsequent chapters.

1.2 HISTORICAL NOTES

Rivets were the principal fasteners in the early days of iron and steel, but occasionally bolts of mild steel were used in structures.[1.6,1.8] It had long been known that hot-driven rivets generally produced clamping forces. However, the axial force was not controlled and varied substantially. Therefore, it could not be evaluated for design.

Batho and Bateman were the first to suggest that high-strength bolts could be used to assemble steel structures.[1.1] In 1934 they reported to the Steel Structures Committee of Scientific and Industrial Research of Great Britain that bolts could

1

be tightened enough to prevent slip in structural joints. It was concluded that bolts with a minimum yield strength of 54 ksi could be tightened sufficiently to give an adequate margin of safety against slippage of the connected parts.

Based on tests performed at the University of Illinois, Wilson and Thomas reported[1.2] in 1938:

> The fatigue strength of high-strength bolts appreciably smaller than the holes in the plates was as great as that of well driven rivets if the nuts were screwed up to give a high tension in the bolt.

Little more was done about high-strength bolting until 1947 when the Research Council on Riveted and Bolted Structural Joints (RCRBSJ) was formed. The purpose of the council, known now as the Research Council on Structural Connections (RCSC), was as follows:

> To carry on investigations as may seem necessary to determine the suitability of various types of joints used in structural frames.

The council sponsored studies on high-strength bolts and rivets and their use in structural connections. The realization that bolts could be extremely useful in the maintenance of bridges helped support developmental work at this early stage. The use of high-strength steel bolts as permanent fasteners has become general since the formation of the RCRBSJ. Prior to that time heat-treated carbon bolts were only used for fitting-up purposes and for carrying the loads during erection. The bolts were tightened to pull the plies of joint material together, but no attempt was made to attain a precise amount of clamping force.

The American Society for Testing and Materials (ASTM) in conjunction with the RCRBSJ prepared a tentative specification for the materials for high-strength bolts, a specification which was first approved in 1949.[1.3] Using the results of research, the RCRBSJ prepared and issued its first specification for structural joints using high-strength bolts in January 1951.[1.4] This specification permitted the rivet to be replaced by a bolt on a one-to-one basis.

In the early 1950s, the installation procedures, the slip resistance of joints having different surface treatments, and the behavior of joints under repeated loadings were studied.[1.6] Outside of the United States high-strength bolts also attracted much attention. Sufficient experience was gained in the laboratory and in bridge construction to enable the German Committee for Structural Steelwork (GCSS) to issue a preliminary code of practice (1956).[1.7] In Great Britain, the general practice was similar to practice and specifications in the United States. The British Standards Institution issued a British Standard (BS) 3139 dealing with bolt material in 1959. In 1960, BS 3294 was issued to establish the design procedure and field practice.

Research developments led to several editions of the RCRBSJ specifications. Allowable stresses were increased, tightening procedures were modified, and new developments such as the use of A490 alloy steel bolts, galvanized joints and bolts, and slotted holes were incorporated.[1.4] The first edition of the *Guide to Design*

Criteria for Bolted and Riveted Connections,[1.13] sponsored by the council and published in 1974, provided a valuable summary of connection and connector behavior for designers and specification writers alike. The presentation of the strength and deformation statements in their most fundamental and basic forms made the guide directly useful for those using the limit states design formats that emerged in the late 1970s.

1.3 TYPES AND MECHANICAL PROPERTIES OF STRUCTURAL FASTENERS

The mechanical fasteners used in structural connections can be classified as either rivets or bolts. Both serve the same purpose, but there are significant differences in appearance. Standards for both types of fasteners are given in Ref. 1.5.

The most commonly used types of structural bolts are (1) the ASTM A307 grade A carbon steel bolt, (2) the ASTM A325 high-strength steel bolt, and (3) the ASTM A490 quenched and tempered alloy steel bolt.[1.3,1.9,1.10]

The ASTM low carbon steel fastener is primarily used in light structures, subjected to static loads. The high-strength A325 and A490 bolts are heavy hex structural bolts used with heavy hex nuts (see Fig. 1.1).

A307 bolts are made of low carbon steel with mechanical properties as designated by ASTM A307. A325 bolts are made by heat-treating, quenching, and tempering medium carbon steel. Two different strength levels are specified, depending on the size of the bolts (see Fig. 1.2).[1.3] The quenched and tempered alloy steel bolt, designated as the A490 bolt, has higher mechanical properties as compared with the A325 high-strength bolt. It was especially developed for use with high-strength steel members. The A490 specification calls for the heavy head and

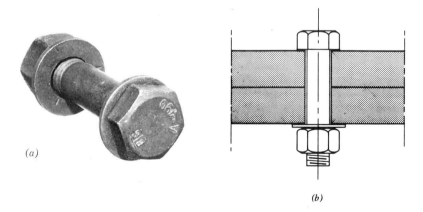

(a)

(b)

Fig. 1.1. Heavy hex bolts. (*a*) High-strength bolt (Courtesy of Bethlehem Steel Corp.); (*b*) installed bolt.

ASTM designation	Bolt diameter (in.)	Tensile strength[a] (ksi)
A307-83a	All	60 minimum
A325-84	$\frac{1}{2}$–1	120 minimum
	$1\frac{1}{8}$–$1\frac{1}{2}$	105 minimum
A490-84	$\frac{1}{2}$–$1\frac{1}{2}$	150 minimum–170 maximum

Fig. 1.2. Tensile strength requirements of structural bolts. [a]Computed on the stress area.

the short thread length of the A325 specification together with chemical and physical properties nearly identical to the A354 grade BD bolt.[1.11] For the development of the A490 bolt many calibration tests were performed on A354 grade BD fasteners manufactured to conform to the A490 specification requirements. The mechanical properties of the different bolt types for structural joints are summarized in Figs. 1.2 and 1.3. Unlike rivets, the strength of bolts is specified in terms of a tensile test of the threaded fastener. The load versus elongation characteristics of a bolt are more significant than the stress versus strain diagram of the parent metal because performance is affected by the presence of the threads. Also, the stress varies along the bolt as a result of the gradual introduction of force from the nut and the change in section from the threaded to the unthreaded portion. The weakest section of any bolt in tension is the threaded portion. The tensile strength of the bolt is usually determined from the "stress area," defined using U.S. Customary Units as:

$$\text{stress area} = 0.785 \left(D - \frac{0.9743}{n} \right)^2$$

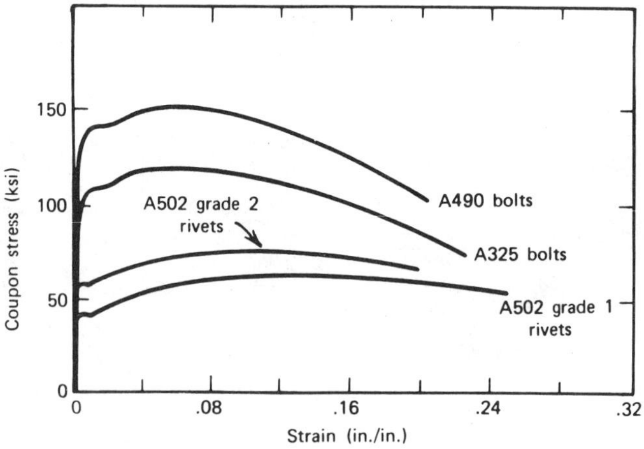

Fig. 1.3. Coupon stress versus strain relationships for different fastener materials.

where D = nominal bolt diameter

       n = number of threads per inch

Figure 1.4 shows typical load versus elongation curves for three different bolts of the same diameter. The tensile strength of each of the bolts was near its specified minimum.

In addition to regular structural bolts, threaded parts have many other structural applications, for example, anchor bolts or tension rods. Anchor bolts are used in column base plates to prevent the uplift of the base plate due to column moments. Threaded parts in tension rods are frequently used to transmit tensile loads from one element to another. In all of these applications, the threaded parts are primarily subjected to tensile loads, and the ultimate tensile load of these connections is determined on the basis of the stress area.

The nut is an important part of the bolt assembly. Nut dimensions and strengths are specified so that the strength of the bolt is developed.[1.5]

Bolts are generally used in holes $\frac{1}{16}$ in. (2 mm) larger than the nominal bolt diameter. When A307 or other mild steel bolts are used, the connection is commonly in bearing, and the nuts are tightened sufficiently to prevent play in the connected members. The clamping force is not very great and should not be considered to have any influence on either the fastener or connection behavior. High-strength bolts (A325 and A490) can produce high and consistent preloads. The RCSC specification requires that they be tightened to at least 70% of the specified minimum tensile strength of the bolt material. The tension is obtained as the nut is turned against the gripped material. Such tightening requires the use of hand torque wrenches or powered impact wrenches. Two methods of controlling bolt

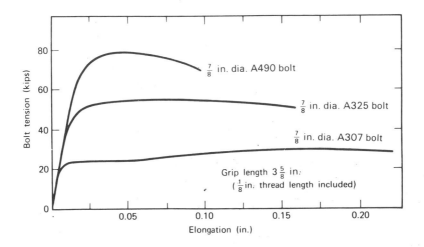

Fig. 1.4. Comparison of bolt types (direct tension).

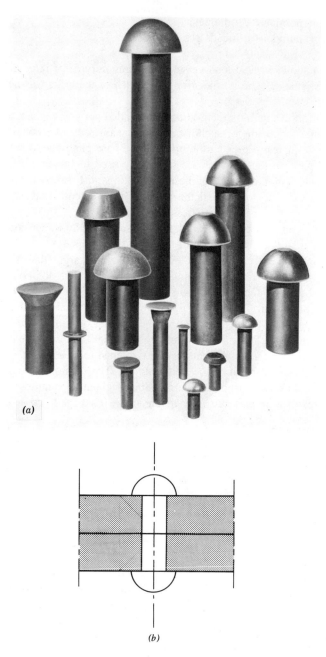

(a)

(b)

Fig. 1.5. Rivets. (a) Rivet types (Courtesy of Bethlehem Steel Corp.); (b) installed rivet.

	Grade 1		Grade 2		Grade 3	
	Min.	Max.	Min.	Max.	Min.	Max.
Rockwell B	55	72	76	85	76	93
Brinell, 500-kgf load, 10-mm ball	103	126	137	163	137	197

Fig. 1.6. Hardness requirements for A502 rivet steel.

tension are used. A detailed description of both tightening procedures is given in Chapter 4.

Rivets are made from bar stock by either hot- or cold-forming the manufactured head. The head is usually of the high button-type, although flattened and countersunk rivets are made for applications with limited clearance. Different rivet types are shown in Fig. 1.5.

Structural rivet steels are of three types: (1) ASTM A502 grade 1, carbon rivet steel, (2) ASTM A502 grade 2, high-strength structural steel rivets, and ASTM A502 grade 3, similar to grade 2 but with enhanced atmospheric corrosion resistance.[1.12] Grade 1 and 2 rivets correspond to those formerly made from steel conforming to ASTM A141 and A195, respectively. Grade 3 rivets are made from steel conforming to ASTM A588. The mechanical hardness requirements for A502 rivet steel are listed in Fig. 1.6. The stress versus strain relationships for typical, undriven A502 rivets are given in Fig. 1.3. For comparative purposes this figure also shows the stress versus strain curves obtained from 0.505-in. diameter specimens turned from full-size A325 and A490 bolts. In North America, rivets are now seldom used in new work. Bolts, either ordinary (A307) or high-strength (A325 and A490), provide an equal or superior connection, and their installed cost is less than that of rivets.

REFERENCES

1.1 C. Batho and E. H. Bateman, *Investigations on Bolts and Bolted Joints*, second report of the Steel Structures Research Committee, London, 1934.

1.2 W. M. Wilson and F. P. Thomas, *Fatigue Tests on Riveted Joints*, Bulletin 302, Engineering Experiment Station, University of Illinois, Urbana, 1938.

1.3 American Society for Testing and Materials, *High-Strength Bolts for Structural Steel Joints*, ASTM Designation A325-84 (originally issued 1949), Philadelphia, 1985.

1.4 Research Council on Riveted and Bolted Structural Joints of the Engineering Foundation, *Specifications for Assembly of Structural Joints Using High-Strength Bolts*, originally issued 1951, latest edition; Research Council on Structural Connections, *Specification for Structural Joints Using ASTM A325 or A490 Bolts*, 1985.

1.5 Industrial Fasteners Institute, *Fastener Standards*, 5th ed., Industrial Fasteners Institute, Cleveland, Ohio, 1970.

1.6 ASCE-Manual 48, Bibliography on Bolted and Riveted Joints, Headquarters of the Society, New York, 1967.

1.7 Deutscher Stahlbau-Verband, *Preliminary Directives for the Calculation, Design and Assembly of Non-Slip Bolted Connections*, Stahlbau Verlag, Cologne, 1956.

1.8 A. E. R. de Jonge, *Bibliography on Riveted Joints*, American Society of Mechanical Engineers, New York, 1945.

1.9 American Society for Testing and Materials, *Heat-Treated Steel Structural Bolts, 150 ksi Minimum Tensile Strength*, ASTM Designation A490-84, Philadelphia, 1985.

1.10 American Society for Testing and Materials, *Carbon Steel Externally Threaded Standard Fasteners*, ASTM Designation A307-83a, Philadelphia, 1983.

1.11 American Society for Testing and Materials, *Quenched and Tempered Alloy Steel Bolts, Studs and Other Externally Threaded Fasteners*, ASTM Designation A354-84b, Philadelphia, 1985.

1.12 American Society for Testing and Materials, *Steel Structural Rivets*, ASTM Designation A502-83a, Philadelphia, 1985.

1.13 J. W. Fisher and J. H. A. Struik, *Guide to Design Criteria for Bolted and Riveted Connections*, Wiley, New York, 1974.

Chapter Two
General Provisions

2.1 STRUCTURAL STEELS

Knowledge of the material properties is a major requirement for the analysis of any structural system. The strength and ductility of a material are two characteristics needed by the designer. These material properties are often described adequately by the stress versus strain relationship for the material. Figure 2.1 shows the stress versus strain relationship that is characteristic of many steels for structural applications. The figure shows the four typical ranges of behavior: the elastic range, the plastic range (during which the material flows at a constant stress), the strain-hardening range, and the range during which necking occurs, terminating in fracture. Generally the initial elastic and yield segments are the most important portions. The following points can be noted in Fig. 2.1:

1. Over an initial range of strain, stress and strain are proportional. The slope of the linear relationship is Young's modulus, E.
2. After the initiation of yield there is a flat plateau. The extent of the yield zone (or ''plastic range'') can be considerable.
3. At the end of the plateau, strain-hardening begins, with a subsequent increase in strength.

Structural steel can undergo sizeable permanent (plastic) deformations before fracture. In contrast to a brittle material, it will generally show signs of distress through permanent, but noncatastrophic, plastic deformation. The energy absorbed during the process of stretching is proportional to the area under the stress versus strain curve. The ductility is essential in various ways for the proper functioning of steel structures and is particularly important in the behavior of connections.

Structural steels can be classified as follows:

1. Structural carbon steel with a specified yield stress between 33 and 36 ksi (228 and 248 MPa). Typical examples are A36 and Fe37 steels.

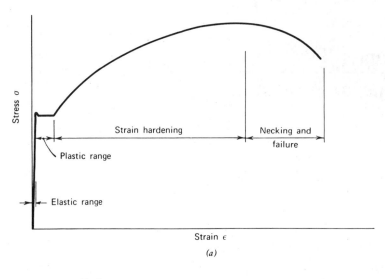

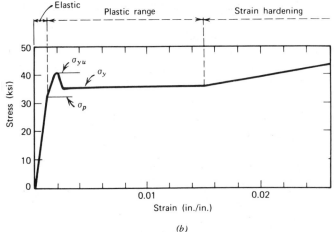

Fig. 2.1. Stress versus strain curve. (*a*) Stress versus strain curve for structural carbon steel; (*b*) initial portion of stress versus strain curve.

2. High-strength steel with a specified yield stress between 42 and 50 ksi (290 and 345 MPa). A typical example in this category is A588 steel.
3. High-strength low-alloy steels with a specified yield stress ranging from 40 to 65 ksi (276 to 448 MPa). This category comprises steels such as A242, A441, A572, A588, and Fe52.
4. Quenched and tempered carbon steel with a specified yield stress between 50 and 60 ksi (345 and 414 MPa). A537 steel is a typical example.

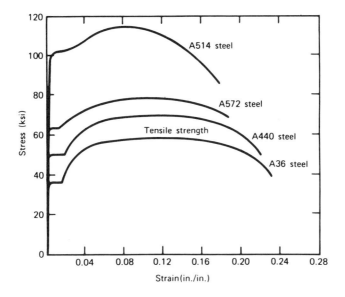

Fig. 2.2. Typical stress versus strain curves for structural steels.

5. Quenched and tempered alloy steel with a specified yield stress ranging from 90 to 100 ksi (621 to 689 MPa). Materials in this category are covered by ASTM A514 and A517.

Typical stress versus strain curves for these steels are given in Fig. 2.2. The curves shown are for steels having specified minimum tensile properties. The corresponding properties of these steels are listed in Fig. 2.3.

Steel Type	Minimum Yield Stress (ksi)	Tensile Strength (ksi)	Minimum Elongation in 8 in.[b] (%)
A36-84a	36	58–80	20
A242-84	42–50[a]	63–70[a]	18
A441-84	40–50[a]	60–70[a]	18
A572-84	42–65[a]	60–80[a]	15–20[a]
A588-84a	42–50[a]	63–70[a]	18
A537-84	45–60[a]	70–100[a]	18–22[a]
A514-84a	90–100[a]	100–130[a]	17–18[a]

Fig. 2.3. Minimum specified properties for structural steels. [a]Depending on thickness. [b]2 in. for A514 steel and for A537 Class 2 steel.

2.2 TYPES OF CONNECTIONS

Mechanically fastened joints are conveniently classified according to the type of forces to which the fasteners are subjected. These classes are (1) shear, (2) tension, and (3) combined tension and shear. Under category 1 the fasteners are loaded either in axial or eccentric shear. If the line of action of the applied load passes through the centroid of the fasteners group, then the fasteners are loaded in axial shear. In eccentric shear the shear force does not pass through the centroid of the fastener group. This results in a torsional moment on the fastener group that increases the fastener shear stresses. This loading condition is referred to as eccentric shear.

The simplest type of structural connection subjecting fasteners to axial shear is the flat plate-type splice. Typical examples are shown in Fig. 2.4a, b, and c. The butt splice is the most commonly used because symmetry of the shear planes prevents bending of the plate material. The load is applied through the centroid of the fastener group. Because two shearing planes cross the fastener, the fasteners act in double shear.

Instead of a symmetric butt splice, the shingle splice (Fig. 2.4b) may be used when the main member consists of several plies of material. A more gradual transfer of load in the plate occurs with this staggered splice than if all main plates are terminated at the same location.

see p 150

Other examples of joints in which the fasteners are subjected to axial shear are gusset plate connections. Depending on the joint geometry, the fasteners are subjected to either double or single shear, as illustrated in Fig. 2.4c. Generally bending is prevented even though the fasteners are in single shear, because of symmetry of the two shearing planes.

In the lap plate splice shown in Fig. 2.4d the fasteners act in single shear. The eccentricity of the loads pulling on the connected members causes bending as the loads tend to align axially. Because of these induced bending stresses, this type of connection is only used for minor connections.

Often situations arise in which the line of the force acting on a connection does not pass through the centroid of the fastener group. This implies that the fastener groups are subjected to eccentric shear forces. Typical examples in this category are bracket connections and web splices of plate girders as shown in Figs. 2.4e and f.

A hanger type connection (Fig. 2.4g) is one of the few examples where mechanical fasteners are used in direct tension. More often the fasteners are subjected to the combined action of tension and shear. This commonly occurs in building frames and bridge deck systems when the connections are required to transmit moments to ensure continuous structural action. The amount of continuity depends on the ability of the connection to resist moments. Moment connections may produce conditions where the upper fasteners are being loaded in shear by the vertical reaction and loaded in tension by the end moment. Some examples of frame connections are given in Fig. 2.4h. (The connectors that may be in combined tension

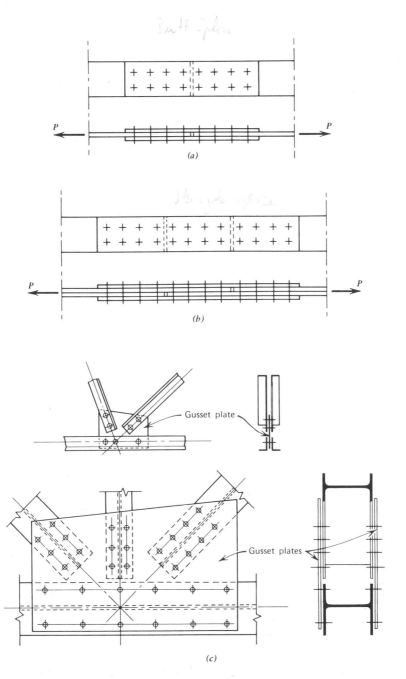

Fig. 2.4. Typical riveted and bolted connections. (*a*) Symmetric butt splice; (*b*) shingle splice; (*c*) single plane construction (top); double plane construction (bottom); (*d*) lap splice; (*e*) bracket connection; (*f*) girder web splice; (*g*) hanger connection. (*h*) frame connection; (*i*) diagonal brace connection.

13

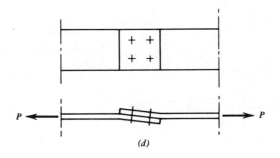

(d)

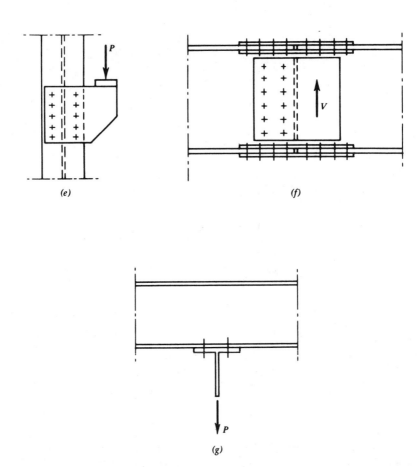

(e)

(f)

(g)

Fig. 2.4. *(Continued)*

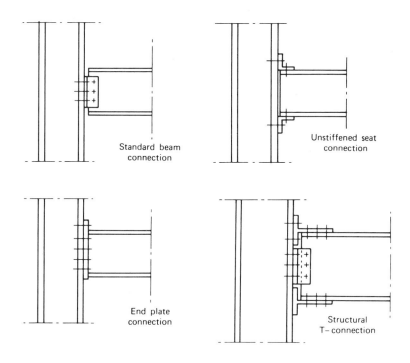

Standard beam connection

Unstiffened seat connection

End plate connection

Structural T-connection

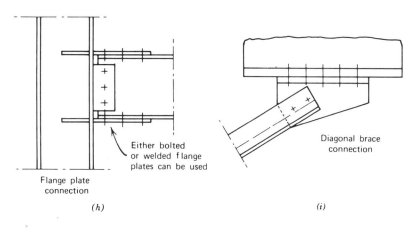

Either bolted or welded flange plates can be used

Flange plate connection

(h)

Diagonal brace connection

(i)

Fig. 2.4. (Continued)

15

and shear are those passing through the column flanges.) Another type of connection in which the fasteners are subjected to combined tension and shear is the diagonal brace shown in Fig. 2.4*i*. (The fasteners that transmit the force from the angle to the web of the tee are in shear only.)

The behavior, analysis, and design of the four major categories of connections—fasteners loaded in axial shear, eccentric shear, tension, or combined tension and shear—are discussed in the following chapters.

2.3 LOADS

The loads and forces acting on a structure may be divided into two broad categories: (1) dead loads and (2) live loads or forces. Dead loads are static, gravitational forces. For a building this usually includes the weight of the permanent equipment and the weight of the fixed components of the building such as floors, beams, girders, and the like. In a bridge it includes the weight of the structural frame, wearing surfaces, lighting fixtures, and such.

As contrasted to the dead loads on a structure, the magnitude of live loads is generally variable with time. Also, most dead loads are static loads, whereas live loads often are at least partially dynamic. In many situations the dynamic nature of the forces has only minor influence on the stress distribution and these loads can be treated as statically applied loads. Live loads can be subdivided into vertical and lateral live loads. The loads on a building due to its occupancy, as well as snow loads on roof surfaces, are regarded as vertical live loads. These load provisions are usually specified in local building codes. Live loads on bridges depend on usage and are specified in the relevant codes such as the AREA[2.1] code for railway bridges and the AASHTO[2.2] specifications that are applicable to highway bridges.

If live loads are dynamic in nature, such as moving vehicles on a bridge or a hoisting machine in a building, it is necessary to account for their dynamic or impact effects. It is well known that the momentum of the load produces internal forces above the static values. In such situations the design load is equal to the sum of the dead load D, and live load L, and the impact load I. The total effect of live load and impact load is usually evaluated by multiplying the live load L by an impact factor p, where p is larger than 1.0. The fraction of p in excess of 1.0 accounts for the load increase because of the dynamic nature of the live load. The impact factor p depends on the type of member, its dimensions, and its loading condition. The factor p is usually prescribed in relevant codes.

Lateral live loads include earth or hydrostatic pressure on the structure and the effects of wind and earthquakes. They also include the centrifugal forces caused by moving loads on curved bridges.

Wind is normally treated as a statically applied pressure, neglecting its dynamic nature. This is justified mainly on the basis of lack of significant periodicity in the fluctuating wind. However, experience has shown this procedure to be unacceptable in certain types of structures, such as suspension bridges and other flexible structures or members. Special consideration of dynamic wind effects is essential in these cases.

An earthquake is a ground motion caused by a sudden fracture and slidings along the fractured surface of the earth crust. Earthquakes are volcanic or tectonic in origin. The forces developed during an earthquake are inertial forces resulting from the tendency of the structure to resist motion. The structure should be capable of resisting these forces with a sufficient margin of safety against distress, that is, full or partial failure or excessive deformations. Some codes, such as the SEAOC[2.3] code, present practical minimum earthquake design procedures for typical structures. In special types of structures a more elaborate analysis of the dynamic response of the structure may be required.

Member forces can also result from temperature effects and support settlements. Consideration must also be given during the design to erection loads.

2.4 FACTOR OF SAFETY—LOAD FACTOR DESIGN

Failure of a structural connection occurs when the externally applied loads exceed the load-carrying capacity (ultimate load). The capacity of a connection can be based on strength or performance criteria. In the first case, loads in excess of the ultimate load lead to a complete or partial collapse of the connection. If performance is the controlling factor, the connection may lose its serviceability before its load capacity is reached because of excessive deformations, fatigue, or fracture. In this respect, unrestricted plastic flow in a structural component is often regarded as determining the useful ultimate load of the member.

Structural members and connections are designed to have a reserve beyond their ordinary service or working load. Allowance must be made for factors such as the variation in quality of materials and fabrication, possible overloads, secondary stresses due to errors introduced by design assumptions, and approximations in calculation procedures. In allowable stress design procedures, a factor of safety is usually employed to provide for these uncertainties. The stress (or load) expected to produce failure is reduced by a factor of safety. This method does not account directly for the statistical nature of the design variables. The expected maxima of loading and the minima of strength not only are treated as representative parameters for design, but also are assumed to occur simultaneously. Neglecting the magnitude and frequency relationships for loads and strengths usually leads to conservative designs. It also results in different reliabilities for the same safety factor.

A different approach to the problem of structural safety can be made by employing the concept of failure probabilities.[2.4, 2.13, 2.17–2.19] Knowing the distributions of the resistance and the load effects, structural safety may be determined from the probability that the effect of the loads will exceed the resistance of the member, as illustrated in Fig. 2.5. The shaded area indicates a finite probability of failure. As the overlap increases, the shaded area, and consequently the failure probability, increases proportionally. Hence, changes in failure probability accompany changes in the load effect-resistance distribution overlap. By employing the

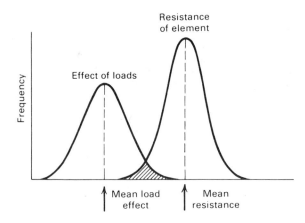

Fig. 2.5.   Probability of failure.

failure probability concept, a uniform reliability throughout the structure can be achieved.

The failure probability of a structural component is considered in a simplified way by the load factor design method.[2.22,2.23] An expression for the maximum strength of a connection can be equated to the strength required to resist the various forces to which it will be subjected. The forces are increased by suitable factors intended to offset uncertainties in their magnitude and application. Thus

$$\phi R = \alpha D + \gamma(L + I)$$

where R represents the average strength, D equals the dead load, and $L + I$ is the summation of the live load and impact load on the connection. The factor ϕ relates to uncertainties in the strength of the connection, whereas the factors α and γ relate to the chance of an increase in load. The factor ϕ is evaluated from a strength distribution curve. The factors α and γ are determined from the distribution curves for dead load and the summation of live load and impact, respectively.

The design recommendations given in the following chapters have been developed considering both the factor of safety concept and the probabilistic approach used in load factor design.

2.5 BOLTED AND RIVETED SHEAR SPLICES

In Section 2.2 different types of connections were classified according to the type of forces to which the fasteners are subjected. If the fasteners in a joint are subjected to shear loads, a further classification based on connection performance is often made. This is illustrated by the behavior of the symmetric butt joint shown in Fig. 2.6. The fasteners can be rivets or bolts, with the clamping force provided by

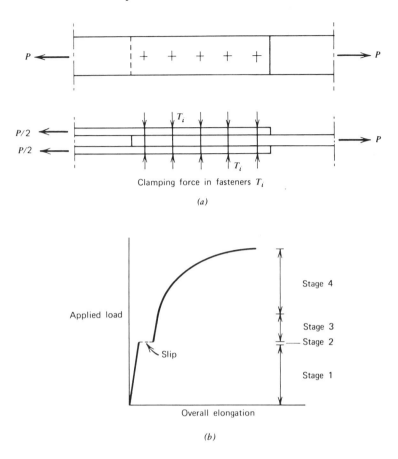

Fig. 2.6. Typical load versus elongation curve of symmetrical butt joint. (*a*) Symmetric butt joint; (*b*) load versus elongation.

tightening the bolts or shrinkage of the rivets due to cooling. If the joint is subjected to an in-plane load through the centroid of the fasteners group, four characteristic loading stages exist, as illustrated in Fig. 2.6. In the first stage, static friction prevents slip; in the second stage, the load has exceeded the frictional resistance, and the joint slips into bearing; in the third stage, the fasteners and plates deform elastically, and consequently the load versus deformation relationship remains linear; in the fourth stage, yielding of plates, fasteners, or both occurs and results in plate fracture or complete shearing of the fasteners. It should be noted that the initial clamping force present in the fasteners has usually been completely dissipated by the time joint failure occurs; the rather small fastener elongations that produced the preload have been released by shear deformation and plate yielding. Overlapping effects may make the distinctions between the various stages less clear-cut than depicted; however, in many tests these stages can be recognized clearly.

In splices subjected to shearing loads, two methods of load transfer are possible: (1) by friction, and (2) by shear and bearing.

If the load on the connection is completely transferred by the frictional resistance on the contact surfaces, it is a slip-resistant joint. Since slip does not occur, these connections are appropriate in situations where slip of the connection is not acceptable, for example in cases involving repeated reversed load conditions or in situations where slip would result in undesirable misalignment of the structure. In slip-resistant joints, the fasteners are not actually stressed in shear, and bearing is not a consideration.

If slip is not considered a critical factor, a load transfer by shear and bearing is acceptable. Depending on the available slip resistance, joint slip may occur before the working load of the connection is reached. Slip brings the connected parts to bear against the sides of the fasteners, and the applied load is then transmitted partially by frictional resistance and partially by shear on the fasteners, depending on joint geometry.

High-strength bolts are very suitable for use in slip-resistant joints, since the magnitude of the axial bolt clamping force, which affects directly the frictional resistance of the connection, can be controlled. This is not true for rivets. Although a clamping force may be developed, it is not reliable. Therefore, riveted joints are usually considered as bearing-type joints.

2.6 FATIGUE

Many structural members may be subjected to frequently repeated cyclical loads. Experience has shown that members and connections under such conditions may eventually fail from fatigue or stable crack growth even though the maximum applied stress is less than the yield stress. In general, fatigue failures occur when the nominal cyclic stress in the member is much lower than the elastic limit. These failures generally show little evidence of deformation. Because of this lack of deformation, fatigue cracks are difficult to detect until substantial crack growth has occurred.

A fatigue fracture surface normally presents a characteristic appearance, with three distinct and recognizable regions. The first region corresponds to slow stable crack growth. This has a visually smooth surface. The second region is rougher in texture as the distance and rate of growth from the nucleus of the fatigue crack increases. The third region is the final fracture, which may be either brittle or ductile, depending on circumstances. Figure 2.7 shows the different stages of a fatigue crack.

For mechanically fastened joints, fatigue crack growth usually starts on the surface at a point of stress concentration such as a hole, a notch, a sharp fillet, a point of fretting, and so on. Notches and other discontinuities cause stress rising effects immediately around the notch and decrease the fatigue strength. The elastic stress concentration factor for an infinitely wide plate with a circular hole and subjected to uniaxial uniform tension is equal to 3.0. Reducing the width of the

Fig. 2.7. Typical fatigue fracture surface.

plate as well as transmitting the load into the plate through a pin-type loading at the hole increases the stress concentration factor significantly. Hence, a change of shape results in a reduction in cross-sectional area and the type of load transfer, both of which are significant factors that influence the magnitude of stress concentrations.

The fatigue life of a particular detail can be obtained either analytically, using fracture mechanics, or by carrying out fatigue tests under controlled conditions. If the fracture mechanics approach is used, a knowledge of both the size and shape of the initial flaw and the stress gradient at the flaw are required; in most civil engineering applications this is not practical. Most of the knowledge of the fatigue life of structural steel members and their connections has been obtained by testing representative details.[2.24] This work has shown that stress range is the dominant stress variable causing crack growth, and that fatigue strength is largely independent of the grade of steel.[2.14, 2.15, 2.21]

It has further been shown that the relationship between stress range ($\Delta\sigma$) and the number of cycles to failure (N) is linear when each variable is expressed in logarithmic form.[2.5-2.7] If sufficient data are available, a mean $\Delta\sigma$-N curve can be determined, as illustrated in Fig. 2.8. This line represents the 50% survival probability of all specimens. The tolerance limits of the $\Delta\sigma$-N curve can be developed

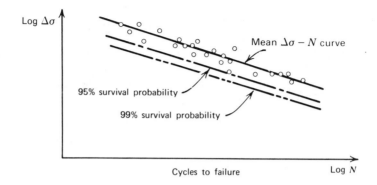

Fig. 2.8. $\Delta\sigma$-N curve and corresponding survival probability curves.

from the variation and survival probability. The desired level of survival probability can be used to develop design stresses for any number of applied stress cycles. Such a procedure is used in Section 5.4 to evaluate design recommendations for bolted joints subjected to repeated loadings.

In recent years the fracture mechanics of stable crack growth has confirmed the suitability of an exponential relationship between cycle life and applied stress range.[2.14,2.15] The tool is of considerable help in evaluating the fatigue behavior of joints.

2.7 FRACTURE

As the temperature decreases, an increase is generally noted in the yield stress and tensile strength of structural steels. In contrast, the ductility usually decreases with a decreasing temperature. Furthermore, there is usually a temperature below which a specimen subjected to tensile stresses may fracture by cleavage; little or no plastic deformation is observed, in contrast to shear failure, which is usually preceded by a considerable amount of plastic deformation. Both types of failure surfaces are shown in Fig. 2.9. Fractures that occur by cleavage are commonly referred to as brittle failures and are characterized by the propagation of cracks at very high velocities. There is little visible evidence of plastic flow, and the fracture surface often appears to be granular except for thin portions along the edges.

Brittle fractures may be initiated at relative low nominal stress levels provided certain other conditions are present, such as (1) a flaw (a fatigue crack or a fabrication crack due to punched holes, etc.), (2) a tensile stress of sufficient intensity to cause a small deformation at the crack or notch tip, and (3) a steel that exhibits low toughness at the service temperature.

To understand brittle fracture one must look at the effects of stress concentrations accompanied by constraints that prevent plastic redistribution of stress. This is the condition that exists in the axially loaded notched bar shown in Fig. 2.10. The

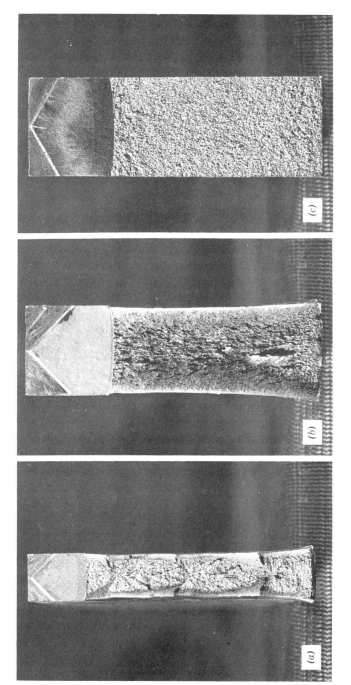

Fig. 2.9. Typical ductile and brittle fracture surfaces. (*a*) Ductile fracture surface with shear lip; (*b*) transition fracture surface; (*c*) brittle fracture surface with flat cleavage surface.

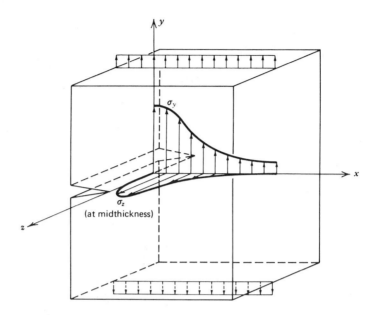

Fig. 2.10. State of stress at the root of a notch under uniaxial loading. Note: σ_y induces σ_z and σ_x stresses. The latter one is not shown in this figure.

stress concentration effect of the notch or crack tip causes high longitudinal stresses at the apex of the crack. These stresses decrease as distance from the apex increases. In accordance with the Poisson effect, lateral contractions must accompany these longitudinal stresses, but the lateral contraction in the width and thickness directions of the highly stressed material at the apex of the notch is restrained by the smaller lateral contractions of the lower stressed material. Consequently, tensile stresses are induced in the width and thickness directions (x and z) so that a severe triaxial state of stress is present near the crack tip. Under these conditions a cleavage- or brittle-type failure may occur.

With decreasing temperatures, the transition from ductile behavior at the crack tip to cleavage behavior occurs within a narrow temperature range. Usually, the Charpy V-notch test is used to evaluate the suspectibility of a steel to brittle fracture. However, in this approach, important factors such as the flaw size and the stress concentration factors are not taken into account. These factors can be accounted for if a fracture analysis or fracture diagram is used.[2.8] The fracture diagram combines fracture mechanics, stress concentration factors, and flaw size with the transition temperature test approach. A detailed description of this concept is given in Refs. 2.8 through 2.10. Considerable work is in progress to assist with the development of fracture mechanics procedures that can be used to define fracture instability conditions. A correlation between the Charpy V-notch and K_{IC}, the plane-strain stress intensity factor at the onset of unstable crack growth, has been suggested.[2.20]

It is apparent that special attention must be directed to design the fabrication details of mechanically fastened connections so that brittle fractures will be avoided. A structural steel with a stable crack growth rate under service conditions should be selected.

One of the critical details in a bolted or riveted structure is the fastener hole. Punching the holes causes strain-aging and work-hardening of the material around the hole. Minute cracks radiating from the hole may form in the work-hardened material, resulting in a notch in a region of high tensile stresses.[2.16]

To eliminate these points of potential crack initiation, holes should either be drilled or subpunched and then reamed in order to eliminate small surface cracks and work-hardened material if brittle fracture is possible under service conditions. Furthermore, geometrical discontinuities such as abrupt changes in cross-section should be avoided.

REFERENCES

2.1 American Railway Engineering Association, *Specifications for Steel Railway Bridges*, Chicago, 1985.

2.2 American Association of State Highway and Transportation Officials, *Standard Specifications for Highway Bridges*, 12th ed., Washington, D.C., 1984.

2.3 Structural Engineers Association of California, *Recommended Lateral Force Requirements and Commentary*, Seismology Committee SEAOC, San Francisco, 1980.

2.4 E. B. Haugen, *Probabilistic Approaches to Design*, Wiley, New York, 1968.

2.5 W. Weibull, *Fatigue Testing and Analysis of Results*, Pergamon, New York, 1961, pp. 174–178, 192–201.

2.6 American Society for Testing and Materials, *A Guide for Fatigue Testing and Statistical Analysis of Fatigue Data*, ASTM Special Technical Publication 91-A, Philadelphia, 1963.

2.7 H. S. Reemsnyder, "Procurement and Analysis of Structural Fatigue Data," *Journal of the Structural Division, ASCE*, Vol. 95, ST7, July 1969.

2.8 W. S. Pellini, "Principles of Fracture Safe Design," *Welding Journal*, Vol. 50, No. 3 and 4, March–April 1971.

2.9 American Society for Testing and Materials, *Fracture Toughness Testing*, ASTM Special Technical Publication 381, Philadelphia, 1965.

2.10 S. T. Rolfe and J. M. Barsom, *Fatigue and Fracture Control in Structures—Applications of Fracture Mechanics*, 2nd ed. Prentice-Hall, Englewood Cliffs, N.J., 1987.

2.11 American Institute of Steel Construction, *Specification for the Design, Fabrication, and Erection of Structural Steel for Buildings*, AISC, Chicago, 1980.

2.12 T. R. Gurney, *Fatigue of Welded Structures*, Cambridge University Press, Cambridge, U.K., 1979.

2.13 J. R. Benjamin and C. A. Cornell, *Probability, Statistics, and Decision for Civil Engineers*, McGraw-Hill, New York, 1970.

2.14 American Society for Testing and Materials, *Fatigue Crack Propagation*, ASTM Special Technical Publication 415, Philadelphia, 1967.

2.15 P. C. Paris, "The Fracture Mechanics Approach to Fatigue," *Proceedings of the 10th Sagamore Conference*, Syracuse University Press, Syracuse, N.Y., 1965, p. 107.

2.16 R. D. Stout, S. S. Tör, and J. M. Ruzek, "The Effect of Fabrication Processes on Steels Used in Pressure Vessels," *Welding Journal*, Vol. 30, September 1951.

2.17 A. M. Freudenthal, "Safety, Reliability and Structural Design," *Transactions ASCE*, Vol. 127, 1962, Part II.

2.18 A. H-S. Ang, "Structural Safety—A Literature Review," *Journal of the Structural Division, ASCE*, Vol. 98, ST4, April, 1972.

2.19 N. C. Lind, "Consistent Partial Safety Factors," *Journal of the Structural Division, ASCE*, Vol. 97, ST6, June 1971.

2.20 J. M. Barsom and S. T. Rolfe, "Correlations between K_{IC} and Charpy V-Notch Test Results in the Transition-Temperature Range," *ASTM STP466, Impact Testing of Metals*, 1970.

2.21 J. W. Fisher, K. H. Frank, M. A. Hirt, and B. M. McNamee, *Effect of Weldments on the Fatigue Strength of Steel Beams*, National Cooperative Highway Research Program Report 102, Washington, D.C., 1970.

2.22 M. K. Ravindra and T. V. Galambos, "Load and Resistance Factor Design for Steel," *Journal of the Structural Division, ASCE*, Vol. 104, ST9, September 1978.

2.23 G. S. Vincent, *Tentative Criteria for Load Factor Design of Steel Highway Bridges*, AISI Bulletin No. 15, American Iron and Steel Institute, Washington, D.C., 1969.

2.24 J. W. Fisher, *Bridge Fatigue Guide—Design and Details*, American Institute of Steel Construction, Chicago, 1977.

Chapter Three
Rivets

3.1 RIVET TYPES

Riveting is among the oldest methods of joining materials, dating back as far as the use of metals in construction practice.[1.8] Rivets were the most popular fasteners during the first half of this century, but their use has declined steadily since the introduction of the high-strength bolts. At the present time they are rarely used in either field or shop connections; either high-strength bolts or welds are used almost exclusively in new work. Nevertheless, the increasing importance of evaluation and retrofitting of existing structures will require that the designer be knowledgeable about riveted connections.

Present specifications (1986) recognize three structural rivet steels, namely ASTM A502 grade 1, carbon rivet steel for general purposes, ASTM A502 grade 2, a carbon-manganese rivet steel suitable for use with high-strength carbon and high-strength low-alloy structural steels, and ASTM A502 grade 3, similar to grade 2 but with enhanced corrosion resistance.[1.12]

It is required that the rivet heads be identified as to grade and manufacturer by means of suitable markings. Markings can either be raised or depressed. For grade 1, the numeral 1 may be used at the manufacturer's option, but it is not required. The use of the numerals 2 or 3 to identify A502 grade 2 or grade 3 rivets is required.

Rivet steel strength is specified in terms of hardness requirements. The hardness requirements are applicable to the rivet bar stock of the full diameter as rolled. Figure 1.6 summarizes the hardness requirements for A502 rivet steels. There are no additional material requirements for strength or hardness in the driven condition.

3.2 INSTALLATION OF RIVETS

The riveting process consists of inserting the rivet in matching holes of the pieces to be joined and subsequently forming a head on the protruding end of the shank. The holes are generally $\frac{1}{16}$ in. greater than the nominal diameter of undriven rivet. The head is formed by rapid forging with a pneumatic hammer or by continuous squeezing with a pressure riveter. The latter process is confined to use in shop practice, whereas pneumatic hammers are used in both shop and field riveting. In addition to forming the head, the diameter of the rivet is increased, resulting in a decreased hole clearance.

Most rivets are installed as hot rivets, that is, the rivet is heated to approximately 1800°F before being installed. Some shop rivets are driven cold, a practice that is permissible if certain procedures are followed.

During the riveting process the enclosed plies are drawn together with installation bolts and by the rivet equipment. As the rivet cools, it shrinks and squeezes the connected plies together. A residual clamping force or internal tension results in the rivet. The magnitude of the residual clamping force depends on the joint stiffness, critical installation conditions such as driving and finishing temperature, as well as the driving pressure. Measurements have shown that hot-driven rivets can develop clamping forces that approach the yield load of a rivet. A considerable variation in clamping forces is generally observed.[3.3,3.6,3.7] Also, as the grip length is increased, the residual clamping force tends to increase.[3.7]

Residual clamping forces are also observed in cold-driven rivets.[3.6] This results mainly from the elastic recovery of the gripped plies after the riveter, which squeezed the plies together during the riveting process, is removed. Generally, the clamping

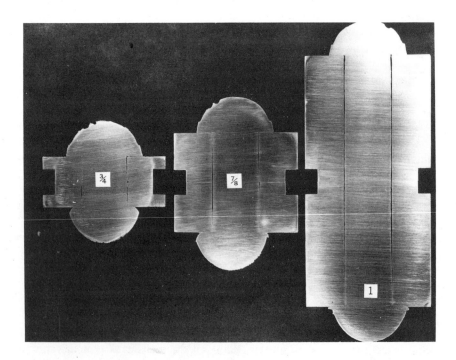

Fig. 3.1. Sawed sections of driven rivets. (Courtesy of University of Illinois.)

force in cold-formed rivets is small when compared with the clamping force in similar hot-driven rivets.

The residual clamping force contributes to the slip resistance of the joint just as do high-strength bolts. However, the clamping force in the rivet is difficult to control, is not as great as that developed by high strength bolts, and cannot be relied upon.

Upon cooling, the rivets shrink diametrically as well as longitudinally. The amount of hole clearance that results also depends on how well the rivet filled the hole prior to shrinkage. Sawed sections of three hot-formed, hand pneumatic driven rivets are shown in Fig. 3.1.[3.2] Studies have indicated that the holes are almost completely filled for relatively short grip rivets. As the grip length is increased, clearances between rivet and plate material tend to increase. This tendency is due to the differences in working the material during driving.[3.2] Figure 3.1 shows some clearance for the longer grip rivets.

Installation of hot-driven rivets involves many variables, such as the initial or driving temperature, driving time, finishing temperature, and driving method. Over the years investigators have studied these factors, and, where appropriate, these results are briefly discussed in the following section.

3.3 BEHAVIOR OF INDIVIDUAL FASTENERS

This section discusses briefly the behavior and strength of a single rivet subjected to either tension, shear, or combined tension and shear. Only typical test data are summarized in this chapter. No attempt is made to provide a comprehensive evaluation and statistical summary of the published test data.

3.3.1 Rivets Subjected to Tension

Typical stress versus strain curves for A502 grade 1 and grade 2 or grade 3 rivet steels are shown in Fig. 1.3. The tensile strength shown in Fig. 1.3 is about 60 ksi for grade 1 and 80 ksi for grade 2 or grade 3 rivets. These are typical of the values expected for undriven rivet materials.

The tensile strength of a driven rivet depends on the mechanical properties of the rivet material before driving and other factors related to the installation process. Studies have been made on the effect of driving temperature on the tensile strength. These tests indicated that varying the driving temperature between 1800 and 2300°F had little effect on the tensile strength.[3.1-3.3] It was also concluded on the basis of these test results that, within practical limits, the soaking time, that is, the heating time of a rivet before driving, had a negligible effect on the ultimate strength.[3.2]

Driving generally increases the strength of rivets. For hot-driven rivets it was observed that machine driving increased the rivet tensile strength by about 20%. The increase was about 10% for rivets driven by a pneumatic hammer. These same increases were observed when the tensile strength was determined from full-size driven rivets and from specimens machined from driven rivets.[3.1] A considerable reduction in elongation was observed to accompany the increase in strength.

Tests also indicated that strain hardening of cold-driven rivets resulted in an increase in strength.[3.1] Although only a few tests are available; they indicate that the increase in strength of cold-driven rivets is at least equal to the increase in strength of similar hot-driven rivets.[3.1,3.4]

Most tension tests of driven rivets showed a tendency to decrease in strength as the grip length was increased. Two factors contribute to this observation. First, there is a greater "upsetting" effect, since the driving energy per unit volume for a short rivet is more favorable. Second, strength figures are based on the full hole area, which implies that the driven rivet completely fills the hole. As was noted in Fig. 3.1, this is not true for longer grip rivets, since the gap increases with increasing grip length.[3.1,3.2] For practical purposes, the differences in strength of short and longer rivets is neglected.

It was reported in Ref. 3.2 that the residual clamping force in driven rivets has no influence on their strength. Yielding of the rivet minimizes the effect of the clamping force and does not affect the ultimate strength. A similar conclusion was reached for preloaded high-strength bolts.[4.5-4.7]

3.3.2 Rivets Subjected to Shear

Many tests have been performed to evaluate the shear capacity of a rivet. It is common practice to express the shear strength of a rivet in terms of its tensile strength.[3.1,3.2,3.5] An average shear strength to tensile strength ratio of about 0.75 has been reported.[3.1,3.2] The grade of the rivet material, as well as whether the test was performed on driven or undriven rivets, had little effect on this average value. Some of the data reported in Refs. 3.1 and 3.2 indicated that the shear to tensile strength ratio varied from 0.67 to 0.83. This wide variation is attributed to differences in testing methods, driving procedures, and test specimens.

Figure 3.2 shows typical load versus deformation curves for double-shear tests on A502 grade 1 rivets.[3.8] Test results of two different grip lengths are shown. As expected, in the initial load stages the longer rivet shows a larger deformation, largely due to bending effects. The shear strength was not affected, however.

Some data indicate a slight decrease in strength for rivets in single shear as

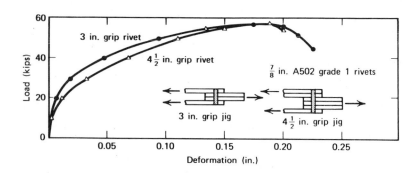

Fig. 3.2. Shear versus deformation curves for A502 grade 1 rivets.

compared with the double shear loading condition. This is caused by out-of-plane forces and secondary stresses on the rivet due to the inherent eccentricity of the applied load. In most single shear test joints, the rivet is not subjected to a pure shear load condition. When a single shear specimen is restrained so that no secondary stresses and out-of-plane deformations are introduced, the difference in the single and double shear strength is insignificant.[3.2]

Since driving a rivet increases its tensile strength, the shear strength is increased as well.[3.1,3.2] If the average tensile strength of undriven A502 grade 1 and A502 grade 2 or grade 3 rivet materials is taken as 60 or 80 ksi, respectively, shear strengths between 45 and 60 ksi for grade 1 rivets and between 65 and 80 ksi for grade 2 or grade 3 rivets can be expected.

3.3.3 Rivets Subjected to Combined Tension and Shear

Tests have been performed to provide information regarding the strength and behavior of single rivets subjected to various combinations of tension and shear.[3.2] ASTM A141 rivets (comparable to A502 grade 1 rivets) were used for the study. The trends observed in this test series are believed to be applicable to other grades of rivets as well.

Among the test variables studied were variations in grip length, rivet diameter, driving procedure, and manufacturing process.[3.2] These variables did not have a significant influence on the results. Only the long grip rivets tended to show a decrease in strength. This was expected and was compatible with rivets subjected to shear alone.

As the loading condition changed from pure tension to pure shear, a significant decrease in deformation capacity was observed. This is illustrated in Fig. 3.3 where

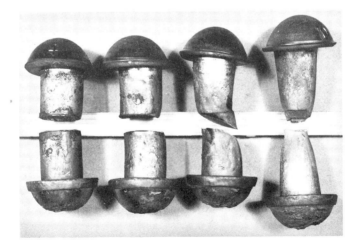

Fig. 3.3. Typical fractures at four shear-tension ratios. (Courtesy of University of Illinois.) Shear-tension ratio: 1.0 : 0.0, 1.0 : 0.577, 0.577 : 1.0, 0.0 : 1.0.

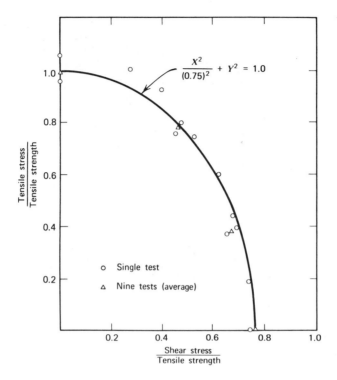

Fig. 3.4.   Interaction curve for rivets under combined tension and shear.

typical fractured rivets are shown for different shear to tension load ratios. Note that the character of the fracture and the deformation capacity changed substantially as the loading condition changed from shear to combined shear and tension and finally to tension.[3.2]

An elliptical interaction curve was fitted to the test results.[3.5] This defined the strength of rivets subjected to a combined tension and shear loading as

$$\frac{x^2}{(0.75)^2} + y^2 = 1.0 \tag{3.1}$$

where x is the ratio of the shear stress on the shear plane to the tensile strength of the rivet (τ/σ_u) and y represents the ratio of the tensile stress to the tensile strength (σ/σ_u). The shear stress and tensile stress were determined on the basis of the applied loads. The tensile strength and shear strength were based on the rivet capacity when subjected to tension or shear only. The test results are compared with the elliptical interaction curve provided by Eq. 3.1 in Fig. 3.4 and show good agreement.

3.4 BASIS FOR DESIGN RECOMMENDATIONS

The behavior of individual rivets subjected to different types of loading conditions forms the basis for design recommendations. This section briefly summarizes rivet strength for the most significant loading conditions.

3.4.1 Rivets Subjected to Tension

The tensile capacity B_u of a rivet is equal to the product of the rivet cross-sectional area A_b and its tensile strength σ_u. The cross section is generally taken as the undriven cross section area of the rivet.[2.11] Hence,

$$B_u = A_b \sigma_u \qquad (3.2)$$

Depending on the type of rivet material, driving method, grip length, and such, σ_u may exceed the undriven rivet strength by 10 to 20%. A reasonable lower bound estimate of the rivet tensile capacity σ_u is 60 ksi for A502 grade 1 rivets and 80 ksi for A502 grade 2 or grade 3 rivets. Since ASTM specifications do not specify tensile capacity, these values can be used.

3.4.2 Rivets Subjected to Shear

The ratio of the shear strength τ_u to the tensile strength σ_u of a rivet was found to be independent of the rivet grade, installation procedure, diameter, and grip length. Tests indicate the ratio to be about 0.75. Hence,

$$\tau_u = 0.75\sigma_u \qquad (3.3)$$

The shear resistance of a rivet is directly proportional to the available shear area and the number of critical shear planes. If a total of m critical shear planes pass through the rivet, the maximum shear resistance S_u of the rivet is equal to

$$S_u = 0.75mA_b\sigma_u \qquad (3.4)$$

where A_b is the cross-section area of undriven rivet.

3.4.3 Rivets Subjected to Combined Tension and Shear

The elliptical interaction curve given by Eq. 3.1 adequately defines the strength of rivets under combined tension and shear (see Fig. 3.4). Equation 3.1 relates the shear stress component to the critical tensile stress component. The product of ultimate stress and the undriven rivet area yields the critical shear and tensile load components for the rivet.

REFERENCES

3.1 L. Schenker, C. G. Salmon, and B. G. Johnston, *Structural Steel Connections*, Department of Civil Engineering, University of Michigan, Ann Arbor, 1954.

3.2 W. H. Munse and H. C. Cox, *The Static Strength of Rivets Subjected to Combined Tension and Shear*, Engineering Experiment Station Bulletin 437, University of Illinois, Urbana, 1956.

3.3 R. A. Hechtman, *A Study of the Effects of Heating and Driving Conditions on Hot-Driven Structural Steels Rivets*, Department of Civil Engineering, University of Illinois, Urbana, 1948.

3.4 W. M. Wilson and W. A. Oliver, *Tension Tests of Rivets*, Engineering Experiment Station Bulletin 210, University of Illinois, Urbana, 1930.

3.5 T. R. Higgins and W. H. Munse, "How Much Combined Stress Can a Rivet Take?," *Engineering News-Record*, Vol. 149, No. 23, Dec. 4, 1952.

3.6 F. Baron and E. W. Larson, Jr., *The Effect of Grip on the Fatigue Strength of Riveted and Bolted Joints*, Research Report C110, The Technological Institute, Northwestern University, Evanston, Illinois, 1952.

3.7 F. Baron and E. W. Larson, Jr., *Comparative Behavior of Bolted and Riveted Joints*, Research Report C109, The Technological Institute, Northwestern University, Evanston, Illinois, 1952.

3.8 J. W. Fisher and N. Yoshida, "Large Bolted and Riveted Shingle Splices," *Journal of the Structural Division, ASCE*, Vol. 96, ST9, September 1969.

Chapter Four
Bolts

4.1 BOLT TYPES

The types of bolts used in connecting structural steel components in buildings and bridges can be categorized as follows (see Section 1.3):

1. Low carbon steel bolts and other fasteners, ASTM A307, grade A
2. High-strength medium carbon steel bolts, ASTM A325, plain finish, weathering steel finish, or galvanized finish
3. Alloy steel bolts, ASTM A490
4. Special types of high strength bolts such as interference body bolts, swedge bolts, and other externally threaded fasteners or nuts with special locking devices, ASTM A449 and ASTM A354 grade BD bolts.

The only marking requirement for ASTM A307 bolts is that the manufacturer's symbol appear on top of the head of the bolt.[1.10] A307 bolts are manufactured with a hexagonal head and nut and either a regular or heavy head, depending on the bolt diameter. Nuts do not need to be marked. The bolts are produced in diameters ranging from $\frac{1}{4}$ to 4 in., have a specified minimum tensile strength of 60 ksi, and may be galvanized.

In application, A307 bolts and nuts are tightened so that some axial force is present that will prevent movement of the connected members in the axial direction of the bolt. Proper tightening also prevents loosening of the nut. The actual force in the bolt is not closely controlled and may vary substantially from bolt to bolt. Because of the small axial forces, little frictional resistance is developed, and in most situations the bolt will slip into bearing. This results in shear stresses in the bolts and contact stresses at the points of bearing.

High-strength bolts are heat treated by quenching and tempering. The most widely used are A325 high-strength carbon steel bolts[1.3] and A490 alloy steel bolts.[1.9]

The A325 bolt is manufactured in diameters ranging from $\frac{1}{2}$ to $1\frac{1}{2}$ in. and is provided as Type 1 (made of medium carbon steel), Type 2 (low-carbon martensite steel), or Type 3 (atmospheric corrosion-resistant steel). Types 1 and 2 can be galvanized. The specified minimum tensile strength for all three types is 120 ksi for bolt diameters up to and including 1 in. and 105 ksi for diameters from $1\frac{1}{8}$ to $1\frac{1}{2}$ in. The bolt heads of all types must be marked A325 and shall also have the manufacturer's symbol. Additional markings distinguish among the three bolt types (see Fig. 4.1). Nut and washer markings are shown in Fig. 4.1. A metric specification is also available for ASTM A325 bolts.[4.30]

Bolts manufactured to ASTM Specification A490 can also be one of three types. Type 1 bolts are made from alloy steel, Type 2 are of low-carbon martensite steel, and Type 3 are of atmospheric corrosion-resistant steel. The bolts are manufactured in diameters ranging from $\frac{1}{2}$ to $1\frac{1}{2}$ in. for all three types, and the specified minimum tensile strength is 150 ksi for all bolts made under this specification. A490 bolts should not be galvanized since they become susceptible to stress corrosion cracking and hydrogen embrittlement.

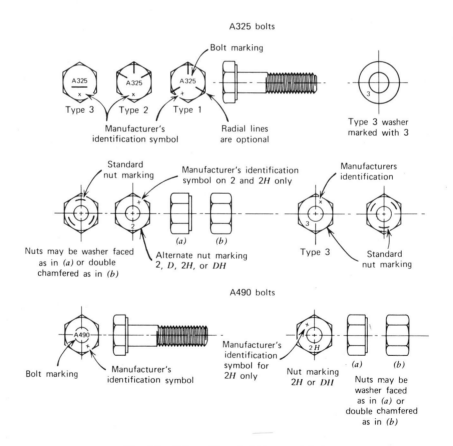

Fig. 4.1. Bolt markings for high-strength bolts.

The markings for A490 bolts are also shown in Fig. 4.1. Bolt heads must be marked with both A490 and the manufacturer's symbol. Other marks, dependent on the bolt type, also appear.

Nuts for A325 bolts must be heavy hex and are required to meet ASTM Specification A563. For bolt Types 1 and 2, plain (uncoated) nut grade C, plain finish, should be used. For bolt Types 1 and 2, galvanized, nut grade DH, galvanized, is required. Nut grade C3 is to be used with bolt Type 3. Grades 2 and 2H nuts, as specified in ASTM A194, and grades D and DH nuts, as specified in ASTM A563, are acceptable alternatives for grade C nuts. Grade 2H nuts (ASTM A194) are an acceptable alternative for grade DH nuts, and type DH3 nuts can be used in place of C3 nuts.

Heavy hex nuts are also required for A490 bolts. Grade DH heavy hex nuts shall be furnished for use with Type 1 and 2 bolts, but grade 2H heavy hex nuts (ASTM A194) are also acceptable. Type 3 A490 bolts require grade DH3 (ASTM A563) heavy hex nuts.

Nuts are marked in various ways, as shown in Fig. 4.1. It should also be noted that both ASTM specifications for high-strength bolts, A325 and A490, stipulate that they are intended for use in structural connections that conform to the RCSC Specification.[1.4]

In addition to the standard A325 and A490 bolts $\frac{1}{2}$ through $1\frac{1}{2}$ in. diameter, short thread heavy head structural bolts above $1\frac{1}{2}$ in. diameter and other types of fasteners and fastener components are available. These are covered by the general bolting specifications A449 and A354. Specification A449 covers externally threaded fastener products with mechanical properties similar to A325. The A354 grade BD covers externally threaded fastener parts that exhibit mechanical properties similar to A490.

Among the special types of fasteners or fastener components are the interference body bolts, swedge bolts, tension-control bolts, and bolt and nut combinations in which the nuts have special locking devices. The interference body bolt (see Fig. 4.2) meets the strength requirements of the A325 bolt and has an axially ribbed shank that develops an interference fit in the hole and prevents excessive slip. A swedge bolt, shown in Fig. 4.3 consists of a fastener pin made from medium carbon steel and a locking collar of low carbon steel. The pin has a series of annular locking grooves, a breakneck groove, and pull grooves. The collar is cylindrical in shape and is swaged into the locking grooves in the tensioned pin by a hydraulically operated driving tool that engages the pull grooves on the pin and applies a tensile force to the fastener. After the collar is fully swaged into the locking grooves, the pin tail section breaks at the breakneck groove when its preload capacity is reached.

Like the swedge bolt, the tension-control bolt is installed by working from one side only, and only one person is required to install the bolt. A special wrench contains a two-part socket that both turns the nut and holds the bolt by means of a splined bolt end. The spline is present toward the end of an extension of the bolt shaft beyond the nut end. This extension also contains a circular notch ("torque control groove") that is calibrated to shear at a torque that will ensure that the

Fig. 4.2. Interference body bolt. (Courtesy of Bethlehem Steel Corp.)

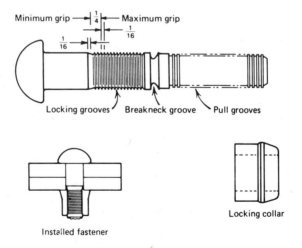

Minimum grip — $\frac{1}{4}$ — Maximum grip

$\frac{1}{16}$

$\frac{1}{16}$ — 11

Locking grooves — Breakneck groove — Pull grooves

Installed fastener

Locking collar

Fig. 4.3. High tensile swedge bolt.

required bolt tension is reached. Installation is quieter than that for a conventional bolt (electric wrenches rather than air-operated impact wrenches are used). Inspection is visual and is simply an observation that the tips have been sheared off. Bolt costs are higher than for conventional high-strength bolts, however, and disposal of the sheared tips may present safety problems.

It should be noted that both swedge bolts and tension-control bolts could be difficult to remove in situations where a structure was being altered or dismantled because they use a rounded head on the bolt.

4.2 BEHAVIOR OF INDIVIDUAL FASTENERS

Connections are generally classified according to the manner of stressing the fastener (see Section 2.2), that is, tension, shear or combined tension and shear. Typical examples of connections subjecting fasteners to shear are splices and gusset plates in trusses. Bolts in tension are common in hanger connections and in beam-to-column connections. Some beam-to-column connections may also subject the bolts to combined tension and shear. It is apparent that, before a connection can be analyzed, the behavior of the component parts of the connection must be known. Therefore, the behavior of a single bolt subjected to the typical loading conditions of tension, shear, or combined tension and shear is discussed in this section.

4.2.1 Bolts Subjected to Tension

Since the behavior of a bolt subjected to an axial load is governed by the performance of its threaded part, load versus elongation characteristics of a bolt are more significant than the stress versus strain relationship of the fastener metal itself.

In the 1985 ASTM specifications for high-strength bolts, both the minimum tensile strength and proof load are specified.[1.3,1.9] The proof load is about equivalent to the yield strength of the bolt or the load causing 0.2% offset. To determine the actual mechanical properties of a bolt, ASTM requires a direct tension test of most sizes and lengths of full-size bolts. In practice, the bolt preload force is usually introduced by tightening the nut against the resistance of the connected material. As this torque is applied to the nut, the portion not resisted by friction between the nut and the gripped material is transmitted to the bolt and, due to friction between bolt and nut threading, induces torsional stresses into the shank. This tightening procedure results in a combined tension-torsional stress condition in the bolt. Therefore, the load versus elongation relationship observed in a torqued tension test differs from the relationship obtained from a direct tension test. Specifically, torquing a bolt until failure results in a reduction in both ultimate load and ultimate deformation as compared with the corresponding values determined from a direct tension test. Typical load versus elongation curves for direct tension as well as torqued tension tests are shown in Fig. 4.4 for A325 bolts and A490 bolts. In torquing a bolt to failure, a reduction in ultimate strength of between 5 and 25% was experienced in tests on both A325 and A490 bolts.[4.1-4.3] The average reduction

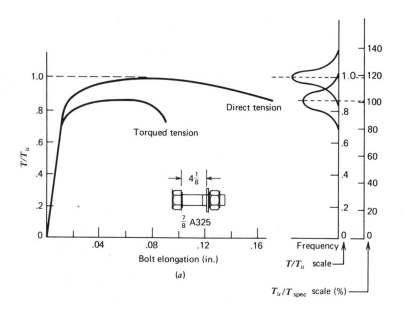

(a)

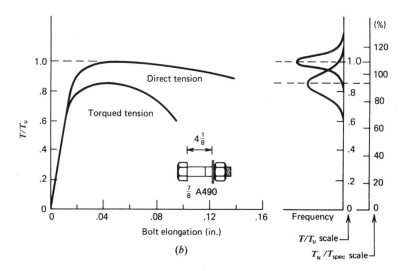

(b)

Fig. 4.4. (a) Load versus elongation relationship and frequency distribution of A325 bolts tested in torqued tension and direct tension; (b) A490 bolts.

is equal to 15%. Frequency distributions of the ratio T/T_u for both A325 and A490 bolts are also shown in Fig. 4.4.

As well as having a higher load, a bolt loaded to failure in direct tension also has more deformation capacity than one that is failed in torqued tension.[4.1-4.3] This is visible in the two specimens shown in Fig. 4.5. The differences in thread deformation and necking of the critical section in the threaded part of the bolts are readily apparent.

To determine whether specified minimum tensile requirements are met, specifications require direct tension tests on full-size bolts if the bolts are longer than three diameters or if the bolt diameter is less than $1\frac{1}{4}$ in. for A325 bolts or 1 in. for A490 bolts. Bolts larger in diameter or shorter in length shall preferably be tested in full size; however, on long bolts tension tests on specimens machined from such bolts are allowed. Bolts shorter than three diameters need only meet minimum and maximum hardness requirements. Tests have illustrated that the actual tensile strength of production bolts exceeds the minimum requirements considerably. An analysis of data obtained from tensile tests on bolts shows that A325 bolts in sizes $\frac{1}{2}$ through 1 in. exceed the minimum tensile strength required by 18%. The standard deviation is equal to 4.5%. For larger diameter A325 bolts (1 to $1\frac{1}{2}$ in.), the range of actual tensile strength exceeds the minimum by an even greater margin. A similar analysis of data obtained from tensile tests on A490 bolts shows an average actual strength 10% greater than the minimum prescribed. The standard variation is equal to 3.5%. Frequency distribution curves of the ratio T_u/T_{spec} are shown in Fig. 4.4a for A325 and in Fig. 4.4b for A490 bolts. Compared with the A325, the A490 bolts show a smaller margin beyond the specified tensile strength because specifications require the actual strength of A490 bolts to be within the range of 150 to 170 ksi, whereas for A325 only a minimum strength is specified.

Loading a bolt in direct tension after having preloaded it by tightening the nut (torqued tension) does not significantly decrease the ultimate tensile strength of the bolt, as illustrated in Figs. 4.6 and 4.7. The torsional stresses induced by torquing the bolt apparently have a negligible effect on the tensile strength of the bolt. This means that bolts installed by torquing can sustain direct tension loads without any apparent reduction in their ultimate tensile strength.[4.1,4.2]

Mean load versus elongation curves for 15 regular head, $\frac{7}{8}$-in. dia. A325 bolts of various grips are plotted in Fig. 4.8.[4.2] The thickness of the gripped material varied from approximately $4\frac{3}{4}$ to $6\frac{3}{4}$ in., and the length of thread under the nut varied from $\frac{3}{4}$ to 1 in. No systematic variation existed among the load versus elongation relationships for the different grip conditions. Most of the deformation occurs in the threaded portion between the underside of the nut and the unthreaded part of the bolt. Because this length is relatively constant, the grip length has no appreciable effect on the load versus elongation response. The behavior shown in Fig. 4.8 for the direct tension test was also observed during torqued tension tests. With shorter grip lengths, the effect of bolt length is more pronounced.

Figure 4.8 also shows that, within the elastic range, the elongation increases

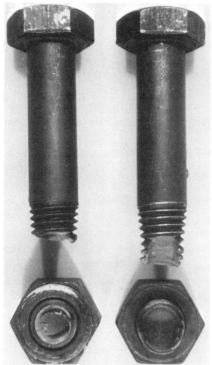

Fig. 4.5. Comparison of torqued tension and direct tension failures.

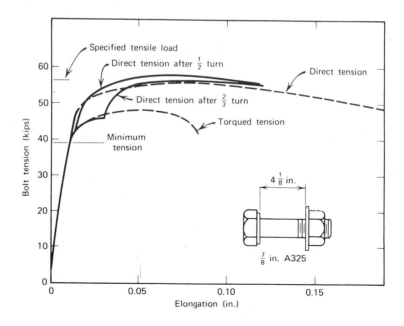

Fig. 4.6. Reserve tensile strength of torqued A325 bolts.

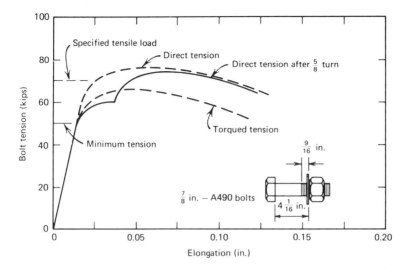

Fig. 4.7. Reserve tensile strength of torqued A490 bolts.

slightly with an increase in grip. As the load is increased beyond the elastic limit, the threaded part, which is approximately of uniform length, behaves plastically, while the shank remains essentially elastic. Hence, when there is a specific amount of thread under the nut, grip length has little effect on the load versus elongation relationship beyond the proportional limit. For short bolts, nearly all deformation occurs in the threaded length, with a resultant decrease in rotational capacity.

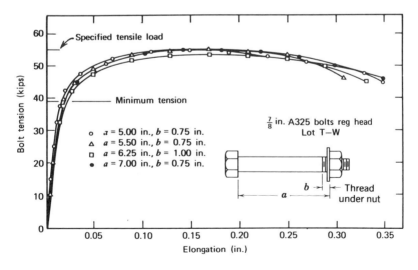

Fig. 4.8. Effect of grip length, direct tension.

A325 bolts with heavy hex heads demonstrate behavior similar to that of bolts with regular heads for grips ranging from 4 to 8 in. and with thread lengths under the nut ranging from $\frac{1}{8}$ to $\frac{3}{4}$ in. Similar observations have also been made about A490 bolts.[4.1,4.3] (Both A325 and A490 bolts are customarily furnished with heavy hexagonal heads unless other dimensional requirements have been agreed on.)

Since most of the elongation occurs in the threads, the length of thread between the thread run-out and the face of the nut will affect the load versus elongation relationship. The heavy head bolt has a short thread length, whereas the regular head bolt has the normal ASA thread length specified by ANSI standards. As a result, for a given thickness of gripped material, the heavy head bolt shows a decrease in deformation capacity, as illustrated in Fig. 4.9.[4.2]

4.2.2   Bolts Subjected to Shear

Shear load versus deformation relationships have been obtained by subjecting fasteners to shear induced by plates either in tension or compression. Typical results of shear tests on A325 and A490 bolts are shown in Fig. 4.10. As expected, the increased tensile strength of A490 bolts as compared with A325 bolts results in an increased shear strength for that fastener. A slight decrease in deformation capacity is evident as the strength of the bolt increases.[4.4]

The shear strength is influenced by the type of test. The fastener can be subjected to shear by plates in tension or compression, as illustrated in Fig. 4.11. The influence of the type of test on the bolt shear and deformation capacity is illustrated in Fig. 4.12, where typical shear stress versus deformation curves are compared for bolts from the same lot that were tested in both tension and compression jigs.[4.4]

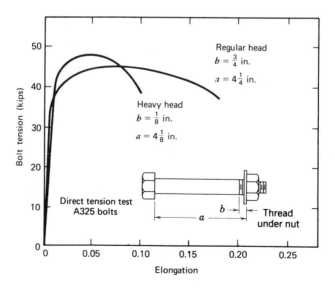

Fig. 4.9.   Comparison of regular and heavy head A325 bolts.

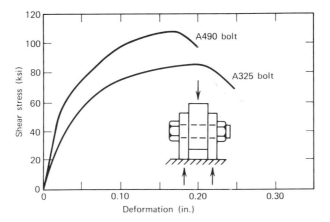

Fig. 4.10. Typical shear load versus deformation curves for A325 and A490 bolts.

Test results show that the shear strength of bolts tested in A440 steel tension jigs is 6 to 13% lower than bolts tested in A440 steel compression jigs.[4.4] The same trend was observed in constructional alloy steel jigs where the reduction in shear strength of similar bolts varied from 8 to 13%. The average shear strengths for A325 and A490* bolts tested in tension jigs were 80.1 and 101.1 ksi, respectively. These shear strengths correspond to about 62% of the respective actual tensile strengths of single bolts. The same bolt grades tested in compression jigs yielded shear strengths of 86.5 and 113.7 ksi, respectively (68% of the bolt tensile strength).[4.4]

The lower shear strength of a bolt observed in a tension type shear test as compared with a compression type test (see Fig. 4.12) is the result of lap plate prying action, a phenomenon that tends to bend the lap plates of the tension jig outward.[4.4,4.25] Because of the uneven bearing deformations of the test bolt, the resisting force does not act at the centerline of the lap plate. This produces a moment that tends to bend the lap plates away from the main plate and thereby causes tensile forces in the bolt.

Catenary action, resulting from bending in bolts, may also contribute to the increase in bolt tension near ultimate load. However, it is believed that this effect is small in comparison with the tension induced by lap plate prying action.[4.25] In any case, the catenary action is present in both the tension and compression jigs.

The tension jig is recommended as the preferred testing device because it produces a lower bound shear strength. Bolts in tension splices are subjected to shear in a similar manner. The tension jig shear test also yields the most consistent test results.

An examination of available test data indicates that the ratio of the shear strength

*Actually, A354 grade BD bolts were used instead of A490 bolts because of their similarity in mechanical properties. At the time of the tests, the A490 bolt was still under development.

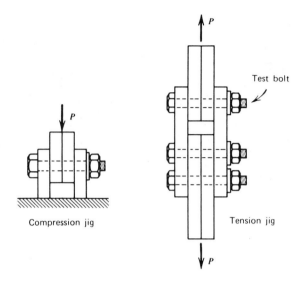

Fig. 4.11. Schematic of testing jigs for single bolts.

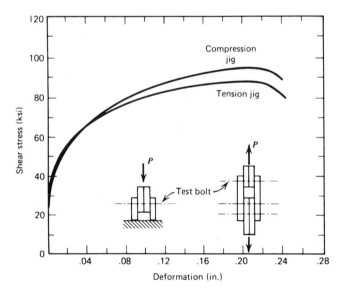

Fig. 4.12. Typical shear load versus deformation curves; A354 BC bolts tested in A440 steel tension and compression jigs.

to the tensile strength is independent of the bolt grade, as illustrated in Fig. 4.13. The shear strength is plotted versus the tensile strength for various lots of A325 and A490 bolts. The average shear strength is approximately 62% of the tensile strength.

The variance of the ratio of the shear strength to tensile strength, as obtained from single bolt tension shear jigs, is shown in Fig. 4.14. A frequency curve of the ratio of shear strength to tensile strength was developed from test data acquired at the University of Illinois and Lehigh University. The average value is equal to 0.62, with a standard deviation of 0.03.

Tests on bolted joints indicated that the initial clamping force had no significant effect on the ultimate shear strength.[4.5-4.7] A number of tests were performed on A325 and A490 bolts torqued to various degrees of tightness and then tested to failure in double shear. The results of tests with A490 bolts are shown in Fig.

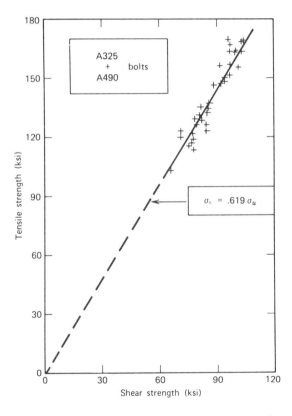

Fig. 4.13. Shear strength versus tensile strength. *Note.* Each point represents the average values of a specific bolt lot. The shear strength is computed on the relevant area, depending on the location of the shear plane.

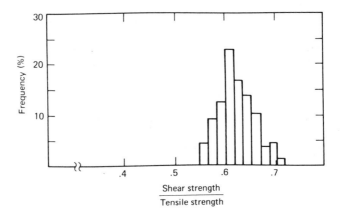

Fig. 4.14. Frequency distribution of ratio of shear strength to tensile strength for A325 and A490 bolts. Number of tests, 142; average value, 0.62; standard deviation, 0.03.

4.15.[4.4] The lower portion shows the relationship between bolt shear strength and the initial bolt elongation after installation. The bolt preload was determined from measured elongations and the torqued tension relationship given in the upper portion of Fig. 4.15. The results confirm that no significant variation of shear strength occurred when the initial bolt preload was varied.

There are two sources of tensile load in the bolt that should, theoretically, interact with the shear load and result in a failure load that is less than that from shear alone. These are (1) the bolt preload induced during the installation procedure, and (2) bolt tension resulting from prying action in the plates.

Measurements of the internal tension in bolts in joints have shown that at ultimate load there is little preload left in the bolt.[4.6,4.7,4.25] The shearing deformations that have taken place in the bolt prior to its failure have the effect of releasing the rather small amount of axial deformation that was used to induce the bolt preload during installation.

At any level of load producing shear in the bolts, prying action of the plates can also produce an axial tensile load in the bolts. In most practical situations, however, the tensile stress induced by prying action will be considerably below the yield stress of the bolt; therefore, it has only a minor influence. Studies of bolts under combined tension and shear have shown that tensile stresses equal to 20 to 30% of the tensile strength do not significantly affect the shear strength of the bolt.[4.8]

The shear resistance of high-strength bolts is directly proportional to the available shear area. The available shear area in the threaded part of a bolt is equal to the root area and is less than the area of the bolt shank. For most commonly used bolts, the root area is about 70% of the nominal area. The influence of the shear plane location on the load versus deformation characteristics of A325 and A490 bolts is reported in Ref. 4.4. Figure 4.16 shows the influence of the shear plane

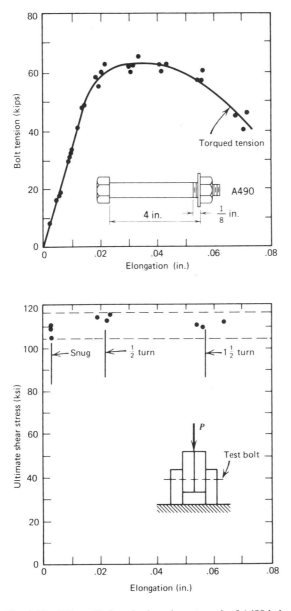

Fig. 4.15.   Effect of bolt preload on shear strength of A490 bolts.

location on the load versus displacement behavior of A325 bolts. When both shear planes passed through the bolt shank, the shear load and deformation capacity were maximized. When both shear planes passed through the threaded portion, the lowest shear load and deformation capacity were obtained. All available tests indicate that the shear resistance of both A325 and A490 bolts is governed by the available

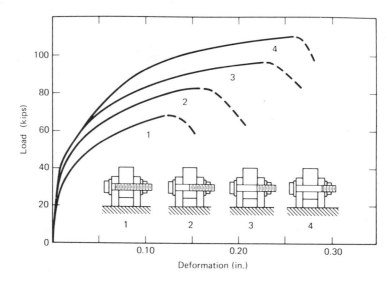

Fig. 4.16. Shear load versus deformation curves for different failure planes.

shear area. The unit shear strength was unaffected by the shear plane location, however.

4.2.3 Bolts Subjected to Combined Tension and Shear

To provide information regarding the strength and behavior characteristics of single high-strength bolts subjected to various combinations of tension and shear, tests were performed at the University of Illinois.[4.8] Two types of high-strength bolts, A325 and A354 grade BD, were used in the investigation. Since the mechanical properties of A354 grade BD and A490 bolts are nearly identical, the data are also directly applicable to A490 bolts.

Certain other factors that might influence the performance of high-strength bolts under combined loadings of tension and shear were also examined in the test program. These included (1) bolt grip length, (2) bolt diameter, (3) type of bolt, and (4) type of material gripped by the bolt. In addition, the influence of the location of the shear planes was examined.

The Illinois tests indicated that an increase in bolt grip tends to increase the ultimate load of a bolt subjected to combined tension and shear. This increase in resistance is mainly caused by the greater bending that can develop in a long bolt as compared with a short grip bolt. At high loads the short grip bolt presented a circular shear area, whereas the long grip bolt, because of bending, presented an elliptical cross-section with a larger shear area.

It was concluded, however, that neither the test block material nor the bolt diameter had a significant effect on the ultimate load capacity of the bolt.

Figure 4.17 summarizes test results of bolts subjected to combined tension and

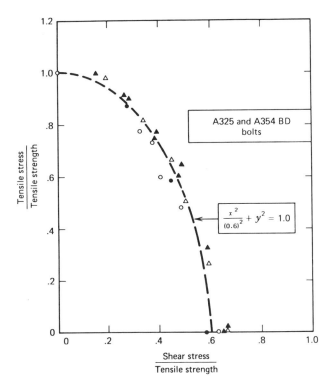

Fig. 4.17. Interaction curve for high-strength bolts under combined tension and shear.

	A325	A354BD
Threads in shear plane	△	○
Shank in shear plane	▲	●

shear.[4.8] The tensile strength (in kilopounds per square inch) was used to nondimensionalize the shear and tensile stresses due to the shear and tensile components of the load. The tensile stress was computed on the basis of the stress area, whereas the shear stress is dependent on the location of the shear plane. An elliptical interaction curve can be used to provide a good representation of the behavior of high-strength bolts under combined tension and shear; namely,

$$\frac{x^2}{(0.62)^2} + y^2 = 1.0 \tag{4.1}$$

where x is the ratio of the shear stress on the shear plane to the tensile strength and y is the ratio of the tensile stress to the tensile strength (both computed on the stress area). Figure 4.17 also indicates that neither the bolt grade nor the location of the shear plane influence the ultimate x/y ratio. This is compatible with the behavior of bolts in pure shear.

4.3 INSTALLATION OF HIGH-STRENGTH BOLTS

North American practice prior to 1985 had been to require that all high-strength bolts be installed so as to provide a high level of preload, regardless of whether it was needed (bolts in a slip-resistant connection or in a connection subject to tension) or not needed (bolts in a bearing-type connection). The advantages in such an arrangement were that a standard bolt installation procedure was provided for all types of connections and that a slightly stiffer structure probably resulted. The disadvantages were economic: the cost of installation of bolts that do not have to be preloaded was increased and the inspection of these installed bolts was unnecessarily complicated.

As was noted in Subsection 4.2.2, the ultimate shear strength of high-strength bolts is not dependent upon the amount of preload in the bolts. There have been a number of specifications that have recognized this in the past,[4.31–4.33] particularly in Europe but also including the International Standards Organization draft specifications for steel structures.[4.34] These specifications permit the use of non-preloaded high-strength bolts in bearing-type connections when load reversals are not present. In 1985, the RCSC introduced a significant relaxation of the rule that had been in previous editions of the specification, namely, that all high-strength fasteners be installed so as to provide a preload equal to 70% of the minimum specified tensile strength of the bolt. The requirement now is that only fasteners that are to be used in slip-critical connections or in connections subject to direct tension need to be preloaded to this level. Bolts to be used in bearing-type connections need only be tightened to the snug-tight condition.

To provide the desired level of preload for bolts used in slip-critical connections or in connections subjected to tension, the RCSC Specification[1.4] continues to require that in these cases the high-strength bolts be tightened such that the resulting bolt tension (preload) is at least 70% of the minimum specified tensile strength of the bolt. The resulting required minimum bolt tension, for various bolt diameters, is given for both A325 and A490 bolts in Table 4.1.

When the high-strength bolt was first introduced, installation was primarily by methods of torque control. Approximate torque values were suggested for use in obtaining the specified minimum bolt tension. For example, early versions of the council specification provided a value of torque that was supposed to produce the required bolt tension (0.0167 lb-ft per inch of bolt diameter per pound tension for standard water-soluble lubricated bolts and nuts). However, tests performed by Maney,[4.12] and later by Pauw and Howard,[4.13] showed the great variability of the torque-tension relationship. Bolts from the same lot yielded extreme values of bolt tension $\pm 30\%$ from the mean tension desired. The average variation was in general $\pm 10\%$. This variance is caused mainly be the variability of the thread conditions, surface conditions under the nut, lubrication, and other factors that cause energy

Table 4.1.   Fastener Tension

Bolt Size (in.)	Minimum Fastener Tension[a] in Thousands of Pounds (kips)	
	A325 Bolts	A490 Bolts
$\frac{1}{2}$	12	15
$\frac{5}{8}$	19	24
$\frac{3}{4}$	28	35
$\frac{7}{8}$	39	49
1	51	64
$1\frac{1}{8}$	56	80
$1\frac{1}{4}$	71	102
$1\frac{3}{8}$	85	121
$1\frac{1}{2}$	103	148

[a] Equal to 70% of specified minimum tensile strengths of bolts, rounded off to the nearest kip.

dissipation without inducing tension in the bolt. Experience in field use of high-strength bolts confirmed the erratic nature of the torque versus tension relationship.

RCSC specifications prior to 1980 permitted high-strength bolts to be tightened by using calibrated wrenches, by the turn-of-nut method, or by use of direct tension indicators.[1.4] The last two procedures depend on strain or displacement control, as contrasted to the torque control of the calibrated wrench method. The 1980 edition of the RCSC specification removed approval for the use of calibrated wrenches, however. (No doubt, many installations were still made using this method. In 1979 it was estimated that about 36% of the bolt installations in the United States were made using calibrated wrenches, but that it was scarcely used at all in Canada.[4.35]) In 1985 the RCSC specification again permitted use of the calibrated wrench method of installation, but with a clearer statement of the requirements of the method and its limitations.

In the calibrated wrench method the wrench is calibrated or adjusted to shut off when the desired torque is reached. In practice, several bolts of the lot to be installed are tightened in a calibrating device that directly reads the tension in the bolt. The wrench is adjusted to shut off at bolt tensions that are a minimum of 5% greater than the required preload. To minimize the variation in friction between the underside of the turned surface and the gripped material, hardened washers must be placed under the element turned in tightening. A minimum of three bolts of each diameter must be tightened at least once each working day in a calibrating device capable of indicating actual bolt tensions. This check must also be performed each time significant changes are made in the equipment or when a significant difference is noted in the surface conditions of the bolts, nuts, or washers.

The calibrated wrench method has a number of drawbacks. Because the method is essentially one of torque control, factors such as friction between the nut and the bolt threads and between the nut and washer are of major importance. The water-soluble lubricant supplied on the bolts can be degraded by rain or moisture or threads can become contaminated with dirt or grease. The result is an erratic torque-tension relationship, and this is not reflected in the calibration procedure. This method of installation also presents field problems when more than one bolt length is used in a given joint because the wrench must be calibrated for each length. (In Japan, the nuts and washers of so-called Quality A high-strength bolt sets are generally treated with a chemical coating in order to overcome some of these problems. The coating reduces the frictional resistance between nut and bolt threads and between nut and washer. However, this coating is sometimes affected by time or temperature.[4.35])

To overcome the variability of torque control, early efforts were made to develop a more reliable tightening procedure. The American Association of Railroads (AAR), faced with the problem of tightening bolts in remote areas without power tools, conducted a large number of tests to determine if the turn-of-nut could be used as a means of controlling bolt tension.[4.14,4.15] These tests led to the conclusion that one turn from a finger-tight position produced the desired bolt tension. In 1955 the RCRBSJ adopted one turn of the nut from hand-tight position as an alternative method to installation.

Experience with the one full turn method indicated that it was impractical to use finger or hand tightness as a reliable point for starting the one turn. Because of out-of-flatness, thread imperfections, and dirt accumulation, it was difficult and time consuming to determine the hand-tight position.

Bethlehem Steel Corporation developed a modified "turn-of-nut" method, using the AAR studies and additional tests of their own.[4.16,4.17] This method called for running the nut up to a snug position using an impact wrench rather than the finger-tight condition. From the snug position the nut was given an additional $\frac{1}{2}$ or $\frac{3}{4}$ turn, depending on the length of the bolt. The snug condition was defined as the point at which the wrench started to impact. This occurred when the turning of the nut was resisted by friction between the face of the nut and the surface of the steel. Snug-tightening the bolts induces small clamping forces in the bolts. In general, at the snug-tight condition the bolt clamping forces can vary considerably because elongations are still within the elastic range. This is illustrated in Fig. 4.18 where the range of bolt clamping force and bolt elongation at the snug tight condition is shown for $\frac{7}{8}$ in. dia. A325 bolts installed in an A440 steel test joint. The average clamping force at the snug-tight condition was equal to about 26 kip. The bolts in this test joint were snug tightened by means of an impact wrench. This modified turn-of-nut method was eventually incorporated into the 1960 specification of the council.

For bolts equal to or greater than about $\frac{3}{4}$ in. dia., snug position provided by an impact wrench is approximately equal to the tightness attained by the full effort of a man using an ordinary spud wrench. For longer or larger diameter bolts, the force produced by this snug load will be less than that for the "standard" case, and for shorter or smaller diameter bolts it will be more. These differences are accommodated in the specification by prescribing the same definition of snug tight

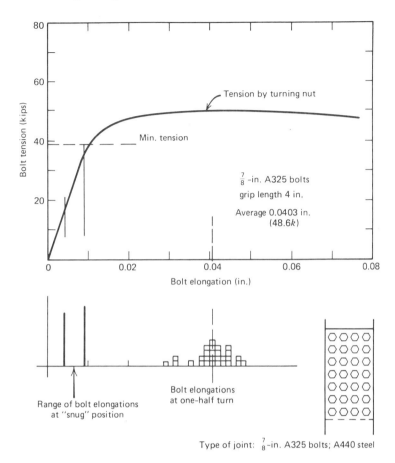

Fig. 4.18. Bolt elongation "snug" and after additional one-half turn of nut. Type of joint: $\frac{7}{8}$ in. dia. A325 bolts; A440 steel.

for all cases but varying the degree of rotation required beyond snug for different situations. As seen in Table 4.2, the current RCSC specification requires one-half turn from snug for bolts whose length from the underside of the head to the extreme end of the bolt is over four but less that eight bolt diameters. If this dimension is less than four bolts diameters, only one-third turn is required, and if it is greater than eight diameters, two-thirds turn is required. Test results are not available for bolts longer than 12 diameters, and so an upper limit is noted in the table. (The definition of bolt length as given previously and in Table 4.2 should not be abused. It is assumed that only a modest projection of bolt beyond the top of the nut will be present. If, for some reason, a large projection is present, the use of Table 4.2 should be based on an adjusted bolt length rather than on the actual bolt length. The length between the underside of the bolt head and the top of the nut would be

Table 4.2. Nut Rotation[a] from Snug-Tight Condition

Bolt Length (as measured from underside of head to extreme end of point)	Disposition of Outer Faces of Bolted Parts		
	Both Faces Normal to Bolt Axis	One Face Normal to Bolt Axis and Other Face Sloped Not More Than 1:20 (bevel washer not used)	Both Faces Sloped Not More Than 1:20 from Normal to Bolt Axis (bevel washers not used)
Up to and including 4 diameters	$\frac{1}{3}$ turn	$\frac{1}{2}$ turn	$\frac{2}{3}$ turn
Over 4 diameters but not exceeding 8 diameters	$\frac{1}{2}$ turn	$\frac{2}{3}$ turn	$\frac{5}{6}$ turn
Over 8 diameters but not exceeding 12 diameters[b]	$\frac{2}{3}$ turn	$\frac{5}{6}$ turn	1 turn

[a]Nut rotation is relative to bolt, regardless of the element (nut or bolt) being turned. For bolts installed by $\frac{1}{2}$ turn and less, the tolerance should be $\pm 30°$; for bolts installed by $\frac{2}{3}$ turn and more, the tolerance should be $\pm 45°$. All material within the grip of the bolt must be steel.

[b]No research work has been performed by the council to establish the turn-of-nut procedure when bolt lengths exceed 12 diameters. Therefore, the required rotation must be determined by actual tests in a suitable tension device simulating the actual conditions.

a suitable choice.) In all cases, care must be exercised to ensure that the nut does not encounter the thread run-out.

Controlling tension by the turn-of-nut method is primarily a strain control. If the elongation of the bolt remains within the elastic range, both the starting point (i.e., snug tight) and the amount and accuracy of the nut rotation beyond snug tight will be influential in determining the preload. However, in the inelastic region the load versus elongation curve is relatively flat, with the consequence that variations in the snug-tight condition result in only minor variations in the preload of the installed bolt. This inelastic behavior will be a characteristic of practically all installed bolts. It results from local yielding of the short length of thread between the underside of the nut and the gripped material. It has no undesirable effect on the subsequent structural performance of the bolt. Figure 4.18 illustrates these points.

Research in the 1960s indicated that one-half turn of the nut from the snug-tight condition was adequate for all lengths of A325 bolts that were then commonly used.[4.2,4.5-4.7,4.9] Based on this experience, the 1962 edition of the council specification required only one-half turn, regardless of bolt length.

In 1964 the council incorporated the A490 bolt into its specification. In order to make the specification applicable to both the A325 and the A490 bolts, the turn-of-nut method was modified again. Tests of A490 bolts had indicated that when the grip length was increased to about eight times the bolt diameter, a somewhat greater nut rotation (two-thirds turn) was needed to reach the required minimum bolt tension. Although the additional rotation was not needed for A325 bolts, the two-thirds turn provision has been applied to the A325 bolts as well in the interest of uniformity in field practice.

Calibration tests of A325 bolts with grips more than 4 diameters or 4 in. showed that the one-half turn of the nut rotation produced consistent bolt tensions in the inelastic range.[4.2] These tests also showed a sufficient margin of safety against fracture by excessive nut rotation. Bolts with grips of more than 4 in. or 4 diameters and short thread length under the nut can be given a one-half turn of the nut and have sufficient deformation capacity to sustain two additional half turns before failure. Bolts with long thread lengths in the grip can sustain three to five additional half turns, as illustrated in Fig. 4.19. Similar tests conducted on A490 bolts allow the comparison with A325 bolts shown in Fig. 4.20. A325 and A490 bolts gave substantially the same load versus nut rotation relationships up to the elastic limit.[4.1,4.3,4.9] At one-half turn from the snug position, the A490 bolts provided approximately 20% greater load than A325 bolts because of the increased strength of the A490 bolt. However, the higher strength of the A490 bolts results in a small decrease in nut rotation capacity as compared with the A325 bolt. These studies show that the factor of safety against twist-off for a bolt installed to one-half turn from snug is about three and one-half for A325 bolts and about two and one-half for A490 bolts. Moreover, it must be recognized that the only source of additional rotation after a bolt is installed would have to be vandalism. Because of the high torque required to produce additional rotation, even this source is unlikely.

Studies on short grip bolts (length less than or equal to four bolt diameters) have shown that their factor of safety against twist-off was less than two when one-half

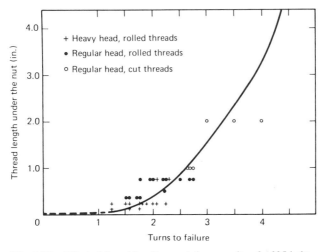

Fig. 4.19. Effect of thread length on rotation capacity of A325 bolts.

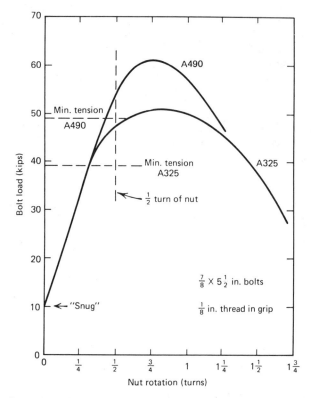

Fig. 4.20. Comparison of bolt load versus nut rotation relationships of A490 and A325 bolts.

turn was used. This resulted in the adoption in 1974 of one-third turn for bolts whose length was less than four diameters. More care needs to be taken in their installation in order to avoid twist-off.

Figure 4.21 shows load versus elongation curves for $\frac{7}{8}$ in. diameter A325 bolts $2\frac{1}{4}$ in. long.[4.36] Some tests were done on low hardness bolts and some on high hardness bolts, and there were either $1\frac{1}{2}$ or $2\frac{1}{2}$ threads unengaged below the nut. It is clear that both parameters had an influence on the ductility of these bolts. High hardness means high strength and reduced ductility. Because most of the bolt elongation is occurring in the threaded portion below the nut, an increase in this length also increased ductility. However, it can be noted that in all cases the specification requirement of one-third turn beyond snug produced a preload greater than the specified minimum value.

It should be apparent that short grip A490 bolts will be potentially less ductile than A325 bolts. Large diameter, short grip bolts will also be of concern because the ratio of tensile stress area to gross area decreases as bolt diameter increases. Figure 4.22 shows unpublished test results on large diameter, short grip A490

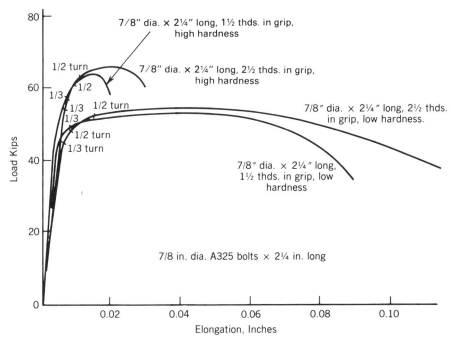

Fig. 4.21. Bolt load versus elongation for short grip A325 bolts.

bolts.[4.37] Because of the relatively large length of unengaged thread below the nut ($\frac{7}{8}$ in.), these bolts showed reasonable ductility for both low hardness and high hardness cases. However, for the same reason, one-third turn beyond snug was not sufficient to produce the specified preload in the bolts. Users of large diameter high-strength bolts, especially A490 bolts, should be aware that the RCSC specification requirement for installation of short grip bolts may not produce the required preload. If such bolts are to be used in a slip-resistant joint, calibration tests in a load-indicating device are advisable.

The 1985 specification of the RCSC permits alternate design bolts to be used and the installation of standard high-strength bolts by means of load-indicating washers. Alternate design bolts include the swedge bolts and tension-control bolts described in Section 4.1. The most common direct tension indicator used is a special washer with projections arranged circumferentially on one flat face. The gap created between the projections and the surface of the steel being clamped will be closed as the preload is introduced into the bolt. Measurements of the size of the gap can be related to the preload. Because of the time required to measure the gap, only spot measurements are usually taken, and care must be exercised to ensure that the protrusions bear against a hardened surface and do not turn as the nut is turned onto the bolt.

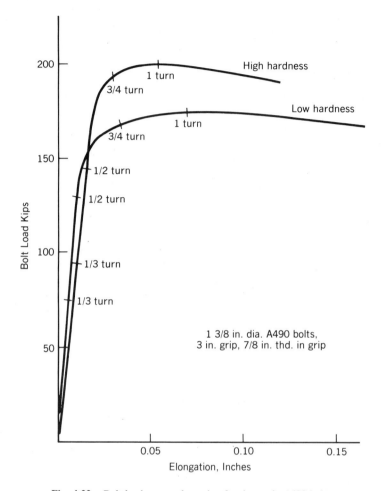

Fig. 4.22. Bolt load versus elongation for short grip A490 bolts.

The council specification contains requirements for verification of load if either alternate design bolts or direct tension indicators are used. Tension-control bolts (see Section 4.1) should be included in this category.

The calibrated wrench method of bolt installation is the most common method used in Japan.[4.35] As an alternative, some installations are made by a method that detects first yield of the bolt threads. An electrically operated wrench using a direct current motor is used. Because the torque in a direct current motor is directly related to the current drawn, it can be used to monitor the bolt tension. A monitor in the wrench is used to detect the first nonlinearity of the operation, and further tightening is prevented. Since the specified minimum tension is below the onset of first yield, this method is satisfactory.

Specifications require that the slope of surfaces of bolted parts in contact with the bolt head or nut shall not exceed 1:20 with respect to a plane normal to the bolt axis. Research carried out at the University of Illinois determined the influence of beveled surfaces (1:20 slope) when bevel washers were omitted.[4.9] A325 bolts are ductile enough to deform to this slope. Greater slopes are undesirable since they affect both strength and ductility.

From these tests it was concluded that the inclusion of bolted connections with a 1:20 slope in the grip and without beveled washers requires additional nut rotation to ensure that tightening will achieve the required minimum tension.[4.9] If one face is normal to the longitudinal axis of the bolt but the other has a bevel of up to 1:20, the usual one-half turn should be increased to two-thirds turn. If both faces are sloped at 1:20, five-sixths turn should be used. Table 4.2 shows the amount of nut rotation for shorter and longer grips when bevelled surfaces are present. Of course, bevel washers can be used to eliminate the slopes and thereby also eliminate the need for additional turns above the standard cases.

4.4 RELAXATION

Because of the high stress level in the threaded part of an installed bolt, some relaxation will occur that could affect the bolt performance. To evaluate the influence of this relaxation, studies were performed on assemblies of A325 and A354 grade BD bolts in A7 steel.[4.9] The bolts were tightened by turning the nut against the gripped material. The bolt tension versus time was registered throughout the study.

From these tests it was evident that immediately upon completion of the torquing there was a 2 to 11% drop in load. The average loss was 5% of the maximum registered bolt tension. This drop in bolt tension is believed to result from the elastic recovery that takes place when the wrench is removed. Creep and yielding in the bolt due to the high stress level at the root of the threads might result in a minor relaxation as well.

The grip length as well as the number of plies are believed to be among the factors that influence the amount of bolt relaxation. Although no experimental data are available, it seems reasonable to expect an increase in bolt force relaxation as the grip length is decreased. Similarly, increasing the number of plies for a constant grip length might lead to an increase in bolt relaxation. Relatively large losses in bolt preload have been reported for very short grip (i.e., $\frac{1}{2}$ to 1 in. grip) galvanized bolts.

Relaxation tests on A325 and A354 BD bolts showed an additional 4% loss in bolt tension after 21 days as compared with the bolt tension measured 1 min after torquing.[4.9] Ninety percent of this loss occurred during the first day. During the remaining 20 days the rate of change in bolt load decreased in an exponential manner.

Relaxation studies on assemblies with high-strength bolts were performed in

Japan and showed similar results.[4.10] By extrapolating the test data, it was concluded that the relaxation after 100,000 hr (11.4 years) could be estimated at about 6% of the bolt load immediately after tightening.

The relaxation characteristics of assemblies of galvanized plates and bolts were found to be about twice as great as plain bolts and connected material.[4.19] The amount of relaxation appeared to be related to the thickness of the galvanized coating. It was concluded that the increased bolt relaxation occurred because of the creep or flow of the zinc coating under sustained high clamping pressures. As with plain ungalvanized bolts, the galvanized bolts experienced most of the creep and relaxation immediately upon completion of the tightening process.

Based on tests performed at Lehigh University, it was concluded that, within certain limits, oversize or slotted holes do not significantly affect the losses in bolt tension with time following installation.[4.26] The loss in tension was about 8% of the initial preload. A more detailed discussion on this is given in Chapter 9.

4.5 REUSE OF HIGH-STRENGTH BOLTS

Since the turn-of-nut method is likely to induce a bolt tension that exceeds the elastic limit of the threaded portion, repeated tightening of high-strength bolts may be undesirable. Tests were performed to examine the behavior of high-strength bolts after torquing one-half turn, loosening, and then retorquing.[4.1,4.2] The record of one such test on a A325 bolt is summarized in Fig. 4.23. It is apparent that the cumulative plastic deformations caused a decrease in the A325 bolt deformation capacity after each succeeding one-half turn. However, A325 bolts can be reused

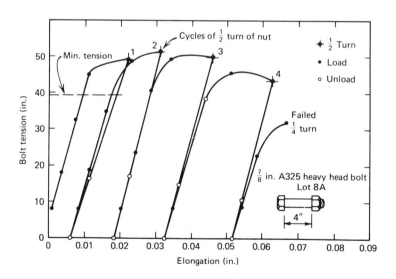

Fig. 4.23. Repeated installation of A325 bolts.

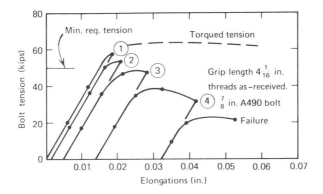

Fig. 4.24. Repeated installation of A490 bolts.

once or twice, providing that proper control on the number of reuses can be established.

As-received high-strength bolts have a light residual coating of oil from the
manufacturing process. This coating is not harmful, and it should not be removed.
Such as-received A325 bolts generally do have adequate nut rotation capacity to
allow for a limited reuse provided either that the original lubricant is still on the
bolt or oil, grease, wax, and so on is applied subsequently. Reuse of coated A325
bolts is not recommended, however. Tests have indicated that the nut rotation
capacity of a bolt is generally reduced by providing a coating (see Section
4.6).[4.19,4.27] Therefore, unless experimental data indicate otherwise, reuse of coated
A325 bolts should not be permitted.

Figure 4.24 shows typical results of one lot of A490 bolts repeatedly installed
with threads as-received. Note that the minimum required tension was achieved
only during the first and second cycle. Subsequent cycles showed a sharp decrease
in induced bolt tension. Test results have indicated that bolts from the same lot
when waxed had considerably improved characteristics.[4.1] However, whether the
threads were waxed or as-received, a marked increase in installation time was noted
for successive cycles. The behavior of A490 bolts under repeated torquing seems
to be more critical than A325 bolts. Therefore, reuse of A490 bolts is not recommended.

4.6 GALVANIZED BOLTS AND NUTS

At the present time, a wide range of structures are being treated with a protective
surface coating to prevent corrosion and reduce maintenance costs. Galvanizing is
a widely used procedure and provides an excellent corrosion-resistant protection.

The behavior of galvanized bolts may differ from the behavior of normal, uncoated high-strength bolts.[4.18,4.19] This difference in behavior is caused primarily
by the zinc layer on the bolt threads. Galling of this zinc layer may take place,

and the nut may seize when the bolt is tightened. Occasionally this makes it difficult to reach the desired bolt tension without experiencing a premature torsional failure of the bolt.

The zinc coating on the surface of a bolt does not affect the bolt static strength properties. Calibration studies showed that neither the tensile strength, as determined from a direct tension test, nor the shear strength of the bolt were affected by the galvanizing process.[4.18,4.19] However, if bolt tension is induced by turning the nut against the gripped material, because of seizure unlubricated galvanized bolts experienced a greater reduction in the maximum bolt tension as compared with torqued ungalvanized bolts or properly lubricated galvanized bolts. This reduction was up to 25% more than that for plain black bolts, depending on the thread conditions and thickness of the zinc layer.

Besides this reduction in torqued tension strength, the added frictional resistance on the threads of the galvanized bolts caused a considerable decrease in ductility, as illustrated in Fig. 4.25. This effect of high frictional resistance can be reduced substantially by employing lubricants on the threads of galvanized bolts. Tests indicated no appreciable difference in the torqued tensile strength of plain bolts as-received and galvanized bolts lubricated with either beeswax, cetyl alcohol, or commercial wax.[4.11,4.27] Some reduction in ductility of the galvanized bolts was observed. Calibration tests performed on galvanized A490 bolts showed results similar to the results of A325 bolts.[4.18]

A high tendency for stripping-type failures was observed in torqued tension tests of galvanized high-strength bolts.[4.19] This can be attributed to several factors. As

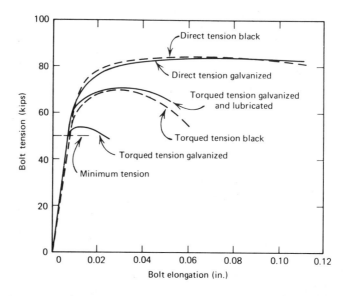

Fig. 4.25. Comparison of bolt load versus elongation relationships between 1-in. black and galvanized A325 bolts.

the bolt is torqued, the threaded section within the grip necks down and the nut spreads. This, along with the overtapping of the nut that is necessary when galvanizing, may cause a disengagement of some of the threads in the nut and increase the chance for stripping failures. To reduce the possibility of an undesirable stripping failure, harder nuts should be used for galvanized bolts (nuts of quality DH or 2H). In order to ensure that the galvanized bolt and nut and the lubricant provided on the nut or bolt threads will provide bolt preload in excess of the specified minimum tension with rotational reserve, special tests are required by the ASTM A325 specification. It must be demonstrated that the galvanized assembly can be subjected to 360° rotation from snug without failure. This test requirement ensures that the tolerances and lubricant are adequate.

Although galvanizing does provide an excellent protection against corrosion of the bolt, it may increase its susceptibility to stress corrosion and hydrogen stress cracking. This applies especially to galvanized A490 bolts. Therefore, it was concluded that galvanized A490 bolts should not be used in structures.[4.23,4.24]

4.7 USE OF WASHERS

Originally the high-strength structural bolt assembly included a bolt with a nut and two hardened washers. The washers were thought necessary to serve the following purposes:

1. To protect the outer surface of the connected material from damage or galling as the bolt or nut was torqued or turned
2. To assist in maintaining a high clamping force in the bolt assembly
3. To provide surfaces of consistent hardness so that the variation in the torque-tension relationship could be minimized

When the turn-of-nut method for tightening high-strength bolts was adopted, a procedure was introduced that provided a means of obtaining the required bolt tension without reliance upon torque-tension control. Hence, it was desirable to determine whether hardened washers were needed in the bolt assembly. Tests showed that a hardened washer was not needed to prevent minor the bolt relaxation resulting from the high stress concentration under the bolt head or nut of A325 bolts.[4.9] It was also concluded that any galling that may take place when nuts for A325 bolts are tightened directly against the connected parts is not detrimental to the static or fatigue strength of the joint.

As a result of these findings, the council specifications in general do not require the use of washers when A325 bolts are installed by the turn-of-nut method. If bolts are tightened by a calibrated wrench method, that is, by torque control, a washer should be used under the turned element, the nut or the bolt head. Washers are required under both the head and nut of A490 bolts when they are used to connect material with a yield point of less than 40 ksi. This prevents galling and brinelling of the connected parts. In high-strength steel they are only required to prevent galling of the turned element.

When bolts pass through a beam or channel flange that has a sloping interface, a bevel washer is often used to compensate for the lack of parallelism. Specifications require the use of beveled washers when an outer face has a slope greater than 1:20. A325 bolts are ductile enough to deform to this slope.[4.9] Greater slopes are undesirable as they affect both strength and ductility.

As noted in Section 4.3, when slopes of up to 1:20 are present in the gripped material, bolts require additional nut rotation to ensure that tightening will achieve the required minimum preload.

There are special requirements for washers when oversize or slotted holes are present; these are described in Chapter 9.

4.8 CORROSION AND EMBRITTLEMENT

Under certain conditions, corrosive environments may be detrimental to the serviceability of coated high-strength bolts subjected to sustained stresses. Hydrogen stress cracking as well as stress corrosion may cause delayed, "brittle" fractures of high-strength bolts. Although both processes have been studied extensively, no completely acceptable mechanism for explaining either phenomenon has been developed.

In many respects the two fracture mechanisms have a number of similarities. Both may cause delayed, brittle-type fractures of bolts. However, there appear to be significant differences. For example, stress corrosion at least in part involves electrochemical dissolution of metal along active sites under the influence of tensile stress. Hydrogen stress cracking occurs as the result of a combination of hydrogen in the metal lattice and tensile stress. The hydrogen produces a hard martensite structure that is susceptible to cracking. Atomic hydrogen absorption by the steel is necessary for this type of failure to occur. Since corrosion frequently is accompanied by the liberation of atomic hydrogen, hydrogen-stress cracking may occur in corrosive environments. However, in many situations a combination of both fracture patterns develops.

The crack surface of a failed bolt that experienced stress corrosion cracking is shown in Figure 4.26. The thumbnail-shaped markings at the bottom of the photograph corresponds to corrosion bands after crack extension and exposure to water. The microscopic appearance of the crack surface near the origin is shown in Figure 4.27 at 2000×. This shows intergranular cracking that is characteristic of stress corrosion cracking.

Laboratory tests have shown that both phenomena influence the life of high-strength bolts.[4.22–4.24] The behavior of A325 as well as A490 bolts under different environmental conditions was studied. From these test results, it became apparent that the higher the strength of the steel, the more sensitive the material becomes to both stress corrosion and hydrogen stress cracking. The study indicated a high

Fig. 4.26. Macroscopic appearance of the crack surface.

susceptibility of galvanized A490 bolts to hydrogen stress cracking. It was concluded that this was caused by a break in the zinc film, which promoted the entry of atomic hydrogen into the metal. If there were no breaks in the coating, failures were not likely to occur. The study also indicated the desirability of limiting the hardness of A490 bolts. Several uncoated bolts were observed to fail when high hardness and strength were present. Because of this observation, the maximum permissible tensile strength was decreased by the ASTM.

On the basis of these tests, it was concluded that properly processed black and galvanized A325 bolts, heat treated within presently specified hardness limits, will behave satisfactorily with regard to hydrogen stress and stress corrosion cracking in most corrosive environments.[4.23] Particular attention should be given to the preparation of the bolts for galvanizing. Improper pickling procedures could induce hydrogen embrittlement. It was further concluded that galvanized A490 bolts should not be used in structures. The tests did indicate that black A490 bolts can be used without problems from "brittle" failures in most environments.

A basic study of the effects of electroplated and hot-dip zinc coatings on the fracture of low-alloy steel AISI 4140 bars in hardness ranges of R_c 33 to 49 was

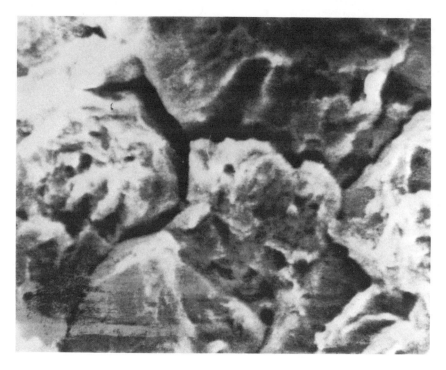

Fig. 4.27. Intergranular fracture surface at the crack origin. 2000×. (Courtesy of Bethlehem Steel Corp.)

conducted by Townsend.[4.38] Electroplated and hot-dip zinc coatings decreased the resistance to stress corrosion cracking directly in relation to the threshold stress intensity, K_{sc}. This effect was attributed to an increased equilibrium hydrogen activity at the crack-tip surface caused by the galvanic effect of the sacrificial coatings. Figure 4.28 shows the measured critical stress intensity as a function of R_c hardness. Although all hardness levels showed stress corrosion susceptibility, the higher hardness levels showed an increased susceptibility.

It was suggested that the condition in bolt threads was directly comparable to the stress intensity for a notched bar, that is,

$$K = \sigma\sqrt{\pi D}\, f\left(\frac{d}{D}\right)$$

For bolts, $f(d/D)$ varies from 0.25 to 0.23 as the shank diameter D varies from $\frac{1}{2}$ to 2 in. and the minor thread diameter d varies from 0.40 to 1.71 in. This approximate fracture mechanics analysis predicts overly conservative results. No failures were observed in actual bolts studied by Boyd and Hyler[4.23] at the lower hardness levels predicted by this study.

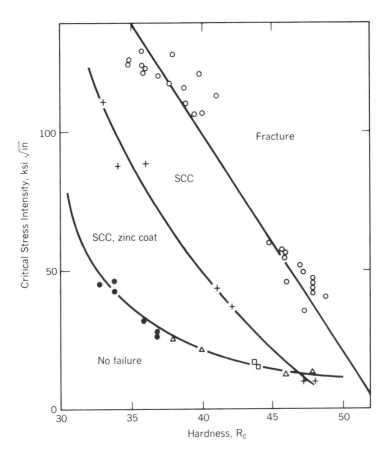

Fig. 4.28. Critical stress intensities in presence of corrosive environment: $\circ$ K_x, all surfaces; $+$ K_{sc}, uncoated; $\bullet$ K_{sc}, hot-dip Al-Zn; $\blacktriangle$ K_{sc}, electroplated zinc; $\blacksquare$ K_{sc}, hot-dip zinc. (From Ref. 4.38)

4.9 EFFECT OF NUT STRENGTH

The behavior of bolt assemblies may vary when tightened to failure. In some cases, failure is in tension through the bolt threads; in other instances, the threads of the nut and/or bolt strip. A tensile failure of the bolt is easily detected; however, a stripping failure develops with imperceptible reduction in torque and is difficult to identify, since some tension remains in the bolt. Therefore, when failure by overtightening occurs or is imminent, a tensile failure of the bolt is preferable. To provide for this, nuts are specified to have a somewhat higher proof load than the bolts with which they are to be used.

As a nut is tightened against the resistance of the gripped material, the bolt lengthens within the grip. If the gripped material and the threads were completely rigid, one turn of the nut would cause the bolt to elongate one pitch. This does not happen, because some thread deformations occur in the bolt and nut. This diminishes the theoretical bolt elongation in the threaded portion.

Since the deformations of the threads are directly affected by the hardness of the nut or the bolt and the number of threads within the depth of the nut, calibration tests were performed on A325 high-strength bolts with minimum and maximum strength levels and assembled with hex nuts and with the thicker heavy hex nuts having various hardness values.[4.20] These tests showed that, with increasing nut hardness, the stripping strength of the connection also increases until the mode of failure changes to a tensile failure in the bolt thread. The bolt tension at one-half turn from a snug-tight condition also increased with an increase in nut hardness, and higher bolt loads were observed in assemblies using high hardness bolts. For all bolt and nut combinations used in this study, the average bolt tension at snug-tight plus one-half turn was considerably above the required minimum tension.

On the basis of these tests,[4.21] as well as other information, the council specification in 1972 started to require the use of heavy hexagonal nuts for A325 and A490 bolts. Because heavy hex nuts have the same dimension across the flats as the bolt head, their use has the additional advantage that a single wrench is applicable to both nut and head.

4.10 BASIS FOR DESIGN RECOMMENDATIONS

The behavior of individual fasteners subjected to different types of loading forms a basis for developing design recommendations. This section summarizes the individual fastener strengths that are used in subsequent chapters to develop design recommendations.

4.10.1 Bolts Subjected to Tension

The tensile capacity of a fastener is equal to the product of the stress area A_s and its tensile strength σ_u. However, it is convenient for design purposes to specify permissible forces and stresses on the basis of the nominal area of the bolt A_b rather than on the stress area A_s. Such a transformation is readily performed because the ratio of the stress area to the nominal bolt area only varies from 0.75 for $\frac{3}{4}$-in. diameter bolts to 0.79 for $1\frac{1}{8}$-in. diameter bolts. The maximum tensile load B_u of a fastener is given as

$$B_u = A_s \sigma_u \qquad (4.2)$$

Expressed in terms of the nominal bolt area and using the lower bound,

$$B_u = 0.75 A_b \sigma_u \qquad (4.3)$$

For most bolt diameters, Eq. 4.3 yields a slightly conservative estimate of the tensile capacity of a bolt.

4.10.2 Bolts Subjected to Shear

The tension-type shear test was observed to provide a lower bound shear strength. The shear strength (in kilopounds per square inch) of a fastener was found to be independent of the bolt grade and equal to 62% of the tensile strength of the bolt material; hence

$$\tau_u = 0.62\sigma_u \tag{4.4}$$

The shear resistance of a bolt is directly proportional to the available shear and the number of shear planes. If a total of m shear planes pass through the bolt shank, the maximum shear resistance S_u of the bolt is equal to

$$S_u = mA_b(0.62)\,\sigma_u \tag{4.5}$$

When shear planes pass through the threaded portion of the bolt, the shear area is equal to the root area of the bolt, which is about 70 to 75% of the nominal bolt area. A lower bound to the maximum shear capacity of the bolt can be expressed as

$$S_u = (0.70)\, mA_b(0.62)\,\sigma_u \tag{4.6}$$

or

$$S_u = (0.43)\, mA_b\sigma_u \tag{4.7}$$

If one shear plane passes through the shank of the bolt and one passes through the threads, the total shear area is equal to the sum of the individual components.

4.10.3 Bolts Subjected to Combined Tension and Shear

An elliptical interaction curve was found to represent adequately the behavior of high-strength bolts under combined tension and shear. The equation (4.1) was given in Section 4.2 as

$$\frac{x^2}{(0.62)^2} + y^2 = 1.0$$

where x is the ratio of the shear stress on the shear plane to the tensile strength and y is the ratio of the tensile stress to the tensile strength (both computed on the stress area). Equation 4.1 relates the shear stress component to the critical tensile stress component. The product of ultimate stress and the appropriate area yields the critical shear and tensile load components.

REFERENCES

4.1 R. J. Christopher, G. L. Kulak, and J. W. Fisher, "Calibration of Alloy Steel Bolts," *Journal of the Structural Division, ASCE*, Vol. 92, ST2, April 1966.

4.2 J. L. Rumpf and J. W. Fisher, "Calibration of A325 Bolts," *Journal of the Structural Division, ASCE*, Vol. 89, ST6, December 1963.

4.3 G. H. Sterling, E. W. J. Troup, E. Chesson, Jr., and J. W. Fisher, "Calibration Tests of A490 High-Strength Bolts," *Journal of the Structural Division, ASCE*, Vol. 91, ST5, October 1965.

4.4 J. J. Wallaert and J. W. Fisher, "Shear Strength of High-Strength Bolts," *Journal of the Structural Division, ASCE*, Vol. 91, ST3, June 1965.

4.5 R. T. Foreman and J. L. Rumpf, "Static Tension Tests of Compact Bolted Joints," *Transactions ASCE*, Vol. 126, Part 2, 1961, pp. 228–254.

4.6 R. A. Bendigo, R. M. Hansen, and J. L Rumpf, "Long Bolted Joints," *Journal of the Structural Division*, Vol. 89, ST6, December 1963.

4.7 J. W. Fisher, P. Ramseier, and L. S. Beedle, "Strength of A440 Steel Joints Fastened with A325 Bolts," *Publications, IABSE*, Vol. 23, 1963.

4.8 E. Chesson, Jr., N. L. Faustino, and W. H. Munse, "High-Strength Bolts Subjected to Tension and Shear," *Journal of the Structural Division, ASCE*, Vol. 91, ST5, October 1965.

4.9 E. Chesson, Jr. and W. H. Munse, *Studies of the Behavior of High-Strength Bolts and Bolted Joints*, Engineering Experiment Bulletin 469, University of Illinois, Urbana, 1965.

4.10 J. Tajima, *"Effect of Relaxation and Creep on the Slip Load of High Strength Bolted Joints,"* Structural Design Office, Japanese National Railways, Tokyo, June 1964.

4.11 P. C. Birkemoe and D. C. Herrschaft, "Bolted Galvanized Bridges—Engineering Acceptance Near," *Civil Engineering*, April 1970.

4.12 G. A. Maney, "Predicting Bolt Tension," *Fasteners*, Vol. 3, No. 5, 1946.

4.13 A. Pauw and L. L. Howard, "Tension Control for High-Strength Structural Bolts," *Proceedings, American Institute of Steel Construction*, April 1955.

4.14 AREA Committee on Iron and Steel Structures, "Use of High-Strength Structural Bolts in Steel Railway Bridges," *American Railway Engineering Association*, Vol. 56, 1955.

4.15 F. P. Drew, "Tightening High-Strength Bolts," Proceeding Paper 786, ASCE, Vol. 81, August 1955.

4.16 E. F. Ball and J. J. Higgins, "Installation and Tightening of High-Strength Bolts," *Transactions, ASCE*, Vol. 126, Part 2, 1961.

4.17 M. H. Frincke, "Turn-of-Nut Method for Tensioning Bolts," *Civil Engineering*, Vol. 28, No. 1, January 1958.

4.18 G. C. Brookhart, I. H. Siddiqi, and D. D. Vasarhelyi, "Surface Treatment of High-Strength Bolted Joints," *Journal of the Structural Division, ASCE*, Vol. 94, ST3, March 1968.

4.19 W. H. Munse, "Structural Behavior of Hot Galvanized Bolted Connections," *Proceedings 8th International Conference on Hot Dip Galvanizing*, London, June 1967.

4.20 E. Chesson, Jr., W. H. Munse, R. L. Dineen, and J. G. Viner, "Performance of Nuts on High-Strength Bolts," *Fasteners*, Vol. 21, No. 3, 1967.

4.21 C. F. Krickenberger, Jr., E. Chesson, Jr., and W. H. Munse, *Evaluation of Nuts for Use with High-Strength Bolts*, Structural Research Series, No. 128, University of Illinois, Urbana, January 1957.

4.22 J. N. Macadam, *Research on Bolt Failures in Wolf-Creek Structural Plate Pipe*, Research Center Armco Steel Corporation, Middletown, Ohio, 1966.

4.23 W. K. Boyd and W. S. Hyler, "Factors Affecting Environmental Performance of High-Strength Bolts," *Journal of the Structural Division, ASCE*, Vol. 99, ST7, July 1973.

4.24 Subcommittee on Bolt Strength, "Delayed Fracture of High-Strength Bolts," *Society of Steel Construction of Japan*, Vol. 6, No. 52, Tokyo, June 1970.

4.25 J. J. Wallaert and J. W. Fisher, "What Happens to Bolt Tension in Large Joints," *Fasteners*, Vol. 20, No. 3, 1965.

4.26 R. N. Allan and J. W. Fisher, "Bolted Joints with Oversize and Slotted Holes," *Journal of the Structural Division, ASCE*, Vol. 94, ST9, September 1968.

4.27 W. H. Munse, *The Case for Bolted Galvanized Bridges*, American Hot Dip Galvanizers Association, Inc., Washington, D. C., May 1971.

4.28 J. H. A. Struik, A. O. Oyeledun, and J. W. Fisher, "Bolt Tension Control with a Direct Tension Indicator," *Engineering Journal, AISC*, Vol. 10, No. 1, 1973.

4.29 W. S. Hyler, K. D. Humphrey, and N. S. Croth, *An Evaluation of the High Tensile Huck-Bolt Fastener for Structural Applications*, Report 72, Huck Manufacturing Co., Detroit, Michigan, March 1961.

4.30 American Society for Testing and Materials, *High-Strength Bolts for Structural and Steel Joints [Metric]*, ASTM Designation A325M-84a, Philadelphia, 1985.

4.31 European Convention for Constructional Steelwork (ECCS), *European Recommendations for Bolted Connections in Structural Steelwork*, 4th Ed., Brussels, November 1983.

4.32 Commission of European Communities (CEC), *Eurocode 3, Common Unified Code of Practice for Steel Structures*, Brussels, November 1983.

4.33 Swiss Society of Engineers and Architects, SIA 161, *Steel Structures*, Zürich, 1979.

4.34 International Standards Organization, *Steel Construction—Materials and Design*, Document ISO/TC 167/SCl N132, Geneva, 1986.

4.35 J. W. Fisher, B. Kato, H. M. Woodward, and K. H. Frank, *Field Installation of High Strength Bolts in North America and Japan*, IABSE Surveys S-8/79, IABSE, Zürich, 1979.

4.36 "Short Grip High-Strength Bolts," unpublished report, University of Illinois, Urbana, May 1979.

4.37 J. W. Fisher, Private communication, April 1984.

4.38 H. E. Townsend, Jr., "Effects of Zinc Coatings on the Stress Corrosion Cracking and Hydrogen Embrittlement of Low-Alloy Steel," Metallurgical Transactions A, Vol. 6A, April 1975.

Chapter Five

Symmetric Butt Splices

5.1 JOINT BEHAVIOR UP TO SLIP

5.1.1 Introduction

A slip-resistant joint (also called a friction-type joint) is one that has a low probability of slip at any time during the life of the structure. It is used where any occurrence of a major slip would endanger the serviceability of the structure and therefore has to be avoided. It should be emphasized that the slip-resistant connection is used to meet a serviceability requirement. Thus, in load factor design, the design of a slip-resistant connection is to be carried out under the working loads, not the factored loads; the joint must not slip in service. (The term "working load" is used throughout this book to represent that load specified by the authority having jurisdiction for the structure. The terms "characteristic load" or "specified load" are often used elsewhere to mean the same thing.)

In a slip-resistant joint, the external applied load usually acts in a plane perpendicular to the bolt axis. The load is completely transmitted by frictional forces acting on the contact area of the plates* fastened by the bolts. This frictional resistance is dependent on the bolt preload and slip resistance of the faying surfaces. The maximum capacity is assumed to have been reached when the frictional resistance is exceeded and overall slip of the joint occurs that brings the plates into bearing against the bolts.

Slip-resistant joints are often used in connections subjected to stress reversals, severe stress fluctuations, or in any situation wherein slippage of the structure into bearing would produce intolerable geometric changes.

In the following sections, the different factors influencing the slip load of a connection are discussed.

5.1.2 Basic Slip Resistance

The slip load of a simple tension splice, as shown in Fig. 5.1, is given by

---

*This term is used here to mean not only plates but any connected parts such as angles, channels, and so on.

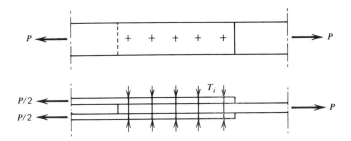

Fig. 5.1. Symmetric shear splice.

$$P_{\text{slip}} = k_s m \sum_{i=1}^{n} T_i \qquad (5.1)$$

where k_s = slip coefficient
               m = number of slip planes
$\sum_{i=1}^{n} T_i$ = sum of the bolt tensions

If the bolt tension can be assumed to be equal in all bolts, this reduces to

$$P_{\text{slip}} = k_s m n T_i \qquad (5.2)$$

where n represents the number of bolts in the joint.

Equation 5.2 shows clearly that for a given number of slip planes and bolts, the slip load of the joint depends on the slip coefficient and bolt clamping force. For a given geometry, the slip load of the connection is proportional to the product of the slip coefficient k_s and bolt tension T_i.

Both the slip coefficient (k_s) and the clamping force (T_i) show considerable variation from their mean values. The slip coefficient varies from joint to joint and, although a specified minimum preload is usually prescribed, bolt preloads are also known to vary considerably, generally exceeding the prescribed minimum value. These variations in the basic parameters describing the slip load must be taken into account when developing criteria for joint design.

5.1.3 Evaluation of Slip Characteristics

The slip coefficient k_s corresponding to the surface condition can only be determined experimentally. In the past, slip tests have usually been performed on symmetric butt joints loaded in tension until slip of the connection occurs. The bolt preload, induced by the tightening process, is determined before the test is started. Once the slip load of the connection is known, the slip coefficient can be evaluated from Eq. 5.2.

$$k_s = \frac{P_{\text{slip}}}{mnT_i} \tag{5.3}$$

Most of the work done to determine the slip coefficient has been on symmetric butt joints of the type shown in Fig. 5.2. Both a two bolt specimen, type A, and a four bolt specimen, type B, have been used. The two standard test specimens with dimensions given in Fig. 5.2 are recommended for use with A325 as well as A490 bolts. Nearly identical specimens have been recommended in Europe by the European Convention for Constructional Steelwork.[5.30] Of course, in fabricating and preparing the test specimens, care must be taken to ensure that the material and surface conditions of the test joints are representative of conditions that occur in the field.

It is apparent from Eq. 5.3 that the value of the bolt clamping force T_i is of prime importance when determining the value of slip coefficient k_s. Since the early stages of high-strength bolting, much attention has been directed to determine the axial force in a bolt installed in a joint. Up to the time of publication, no precise method is available. The best available method is to calibrate the bolts used in the test specimens.[4.2] This requires that each bolt be calibrated prior to installation in the test joint. The bolt clamping force should be within the elastic range if an

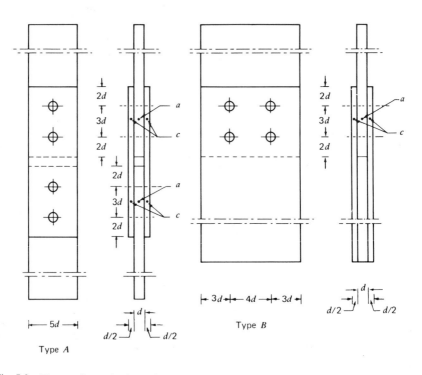

Fig. 5.2. Test specimens for determining the slip coefficient. Bolt diameter, d; hole diameter, $d + \frac{1}{16}$ in.

accurate evaluation is made. Consequently, the bolts can be used more than once as long as the grip length is not altered. If the bolts are tightened beyond the elastic limit load, permanent plastic bolt deformations will occur. In such cases an average bolt load versus elongation curve for the lot to be used in the test joints has to be determined from a representative sample of bolts. The elongations of the bolts in the test joint can be related to the clamping force through this average bolt calibration curve. Because of inelastic deformations, the bolts can only be used once.

As described in Chapter 4, load-indicating devices such as tension control bolts and load-indicating washers are available for establishing that the bolt preload meets or exceeds the specified minimum value. Whether such devices would provide a sufficiently accurate measure of the preload for purposes of the slip test would have to be evaluated on a case-by-case basis.

More recently, the RCSC has developed a standard test for determination of the slip coefficient when coatings are used in bolted joints. The "standard" specimen in this case is a three-plate specimen (one main plate and two lap plates) loaded in compression and containing a single fastener. The fastener described in the test method is actually a threaded rod and nut arrangement that permits application of a known load by means of a center hole ram. Alternative means of applying the clamping force are permitted, including use of a high-strength bolt, as long as the magnitude of the force in the bolt can be established to within $\pm 1\%$. A tension-type test is also permitted, and the specification provides rules for establishing the performance of connections under sustained loads (creep). The slip behavior of bolted joints when coatings are used on the faying surfaces is discussed in Chapter 12.

Regardless of which type of specimen is used to carry out the slip test, in a short-term static test the test specimens are subjected to gradually or incrementally increasing tensile loads. The displacements between points a and c (see Fig. 5.2) should be recorded at selected intervals of loadings.

In most slip tests on specimens without a protective coating on the slip surfaces, a sudden slip occurs when the slip resistance of the connection is exceeded. Coated specimens often do not exhibit sudden slip; the slipping builds up continuously as evidenced by cumulative microslips. In these situations the load corresponding to a prescribed amount of slip, usually 0.02 in., can be used to define the slip load.

Other than major slip, creep of a connection might impair the serviceability of a joint as well. A creep test can be performed to evaluate the influence of sustained loading levels on the displacement of a joint. A constant load level is applied for a long period in a creep test, and the observed displacements are evaluated. The RCSC specification for determination of slip loads can be consulted for details of a suitable creep test.

5.1.4 Effect of Joint Geometry and Number of Faying Surfaces

The effects of joint geometry have been examined in numerous experimental studies. The significance of the influence of factors such as number of bolts in a line and whether the bolts are arranged in compact patterns has not been determined.

An analysis of the slip coefficient in large bolted joints having clean mill scale surfaces yields an average slip coefficient 0.33 with a standard deviation of 0.07. For small joints these values were 0.34 and 0.07, respectively. In this comparison, a large bolted joint was defined as having at least two lines of bolts parallel to the direction of the applied load, with each line consisting of at least three bolts. Based on the results of this analysis, it was concluded that the number of bolts in a joint does not have a significant influence on the slip coefficient.

The slip resistance of a bolted joint is also proportional to the number of faying surfaces. Hence, a multilap joint can resist slip with great efficiency. Tests have shown that the slip coefficient is not affected by the number of faying surfaces, however.[5.34]

5.1.5   Joint Stiffness

In slip-resistant joints the main plate and lap plates are compressed laterally by the initial clamping force. No relative displacement of the contact points on the surfaces takes place, and the joint may be considered equivalent to a solid piece of metal with a cross-section equal to the total area of the main and lap plates.

The stiffness of the joint, characterized by the slope of the load versus deformation curve, will decrease significantly if yielding occurs in either the net or gross cross-section. Yielding will not occur under working load levels because the working load is much less than the yield load of the connection. Since, under either allowable stress design or load factor design, the slip-resistant connection is designed using the working loads, its stiffness will not be affected by yielding up to the load levels for which the design is applicable.

5.1.6   Effect of Type of Steel, Surface Preparation, and Treatment on the Slip Coefficient

One of the significant factors influencing the slip resistance of a connection is the slip coefficient k_s, as defined by Eq. 5.3. Because of its significant influence, much research has been done in the United States, Europe, Japan, and elsewhere to determine the magnitude of k_s for different steels, different surface treatments, and surface conditions.[4.5–4.7, 4.26, 5.1–5.17] The results of these studies have been used to evaluate the slip coefficient for a number of surface conditions.

It is clear that to determine a reliable value of the slip coefficient k_s, an accurate estimate of the initial clamping force must be known. Therefore, only tests where the actual clamping force in the bolts was measured were considered in the following analysis. Data obtained from tests in which bolts were installed using torque control were not considered.

In many cases structural members are bolted together without special treatment of the faying surfaces. A natural faying surface is provided by clean mill scale. Only the loose mill scale and dirt is removed by hand wire brushing. Grease originating from the fabrication process is removed with a solvent. An analysis of the available data shows that the clean mill scale condition for A7, A36, and A440 structural steels yield an average slip coefficient k_s of 0.33, with a standard deviation of 0.06. (Steel manufactured in accordance with ASTM A7 is no longer available, but many of the early test results for slip coefficient were obtained using this steel. The slip characteristics of joints made using A7 steel are considered to be comparable to those obtained using A36 steel.) Tests performed in Europe on Fe37 and Fe52 steels, comparable to A7, A36, and A440 steels, exhibited similar results. If all the available data on A7, A36, Fe37, A440, and Fe52 steel are considered, an average value of k_s equal to 0.33 is obtained, with a standard deviation of 0.07. Fig. 5.3 shows the frequency distribution of the slip coefficient as derived from the 327 tests.

Some slip test results are available for a newer steel, A588, a weathering steel used mainly for bridge structures.[5.55] The data from 31 tests show that the slip coefficient for this steel in the clean mill scale condition is 0.23, with a standard deviation of 0.03. These test results fall on the low side of the scatter shown in

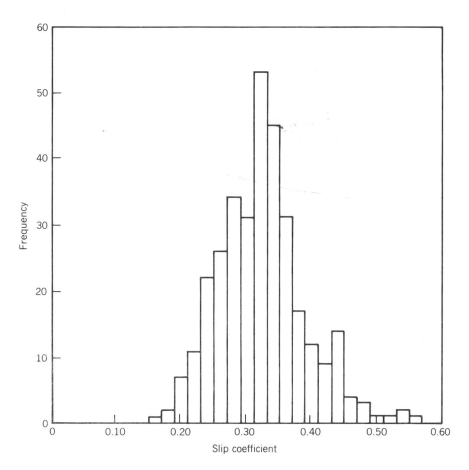

Fig. 5.3. Distribution of slip coefficient for clean mill scale surfaces. Clean mill scale surfaces: A7, A36, A440, Fe37, and Fe52 steel. Number of tests, 327; average, 0.33; standard deviation, 0.07.

Fig. 5.3. However, the results do not differ significantly from other studies contained within Fig. 5.3. For example, Ref. 4.26 reported a mean slip coefficient of 0.25 and a standard deviation of 0.04 for A440 steel specimens. In Ref. 5.15, the slip coefficient reported for A36 steel was 0.27, with a standard deviation of 0.05.

If the mill scale is removed by brushing with a power tool, a shiny clean surface is formed that decreases the slip resistance. Joints tested at Lehigh University with such semipolished faying surfaces indicated a decrease in friction resistance of 25 to 30% as compared with normal hand brushing mill scale surfaces.[5.6] This decrease is mainly due to the polishing effect of the power tool; the surface irregularities, which are essential for providing the frictional resistance, are reduced, causing a decrease in k_s.

Many tests have shown that blast-cleaning with shot or grit greatly increases the slip resistance of most steels as compared with the clean mill scale condition.[5.5,5.11] An analysis of available data yielded an average value k_s equal to 0.51 for A7, A36, and Fe37 steels with blast-cleaned surfaces. The frequency distribution of the test results is shown in Fig. 5.4. It is apparent that the frequency distribution is somewhat skewed. This is reasonable, since the higher values could be influenced by yielding of the steel. The friction coefficient for blast-cleaned A440 and Fe52 steel should not differ from the value reported for blast-cleaned A7, A36, and Fe37 steel surfaces.

The magnitude of k_s for shot-blasted surfaces is greatly affected by the type and condition of grit or material that is employed to clean the surface. The condition of the cleaning material determines whether the surfaces are polished or left with a rough texture that is more slip resistant.

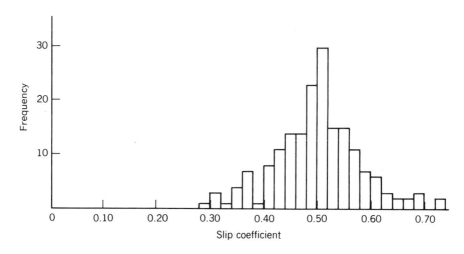

Fig. 5.4. Distribution of slip coefficient for blast-cleaned surfaces. Blast-cleaned surfaces: A7, A36, Fe37 steel; number of tests, 186; average, 0.51; standard deviation, 0.09.

The mean slip coefficients of the three studies contained within Fig. 5.4 varied from 0.49 to 0.55, with standard deviations of between 0.06 and 0.09. A limited study using ST52 steel yielded a mean slip coefficient of 0.65 and a standard deviation of 0.08. Differences in the slip resistance from the different studies may be due to different blast-cleaning procedures in use at the time the studies were undertaken. It should also be noted that the standard deviation of the slip coefficients of the blast-cleaned surfaces does not differ appreciably from the variation observed for clean mill scale surfaces.

Tests on A514 constructional alloy steel showed an average slip coefficient of 0.33 for steel grit-blasted surfaces. Although not much experimental evidence is available, these results show that grit blasting of quenched and tempered alloy steel as compared with lower strength steel has less effect on the slip coefficient. This indicates that the hardness of the surface influences the roughness achieved by the blast cleaning.

In most field situations, structural members are exposed to the atmosphere for a period of time before erection. During this period unprotected blast-cleaned surfaces are highly susceptible to surface corrosion. To simulate this field condition, tests were performed in which the blast-cleaned surfaces were stored in the open air for different periods before being assembled and tested.[5.11,5.15] These test specimens were bolted up without wire brushing or otherwise disturbing the rusted surfaces. The results indicated that the relatively high slip coefficient obtained by shot or grit blasting is decreased with increased exposure time. After 12 months exposure to a humid, industrial atmosphere, the slip coefficient was about the same as the high end of the test results for clean mill scale. Removing the rust by wire brushing improved the slip resistance. If it can be ensured that the blast-cleaned surfaces will be exposed only for a short time, the relatively high slip coefficient of 0.51 (see Table 5.1) can be used for such steels as A36, Fe37, and Fe52.

A distinction must be made in some cases between surfaces blast-cleaned with shot or grit and those cleaned by sand blasting. Quenched and tempered steels, like A514, which have a low coefficient of slip if they have been cleaned using shot, display a much higher coefficient if sand has been used. The test results for sand-blasted A572 and A514 steels can be included with A7 and A36 test results. As seen in Table 5.1, the average slip coefficient for this group is 0.52, with a standard deviation of 0.09.

If rust forming on the blast-cleaned faying surfaces cannot be tolerated, a protective coating can be applied to the surfaces. These protective treatments alter the slip characteristics of bolted joints to varying degrees. Tests have been performed to evaluate the behavior of bolted joints in which the faying surfaces were galvanized, cold zinc painted, metallized, treated with vinyl wash or linseed oil, or treated with rust preventing paint.[5.5,5.9,5.11,5.13,5.18,5.36,5.37] The results of these tests are summarized in Table 5.1. Some of the values listed in this summary were determined from a rather small number of tests. They provide only an indication of the magnitude of the slip coefficient. Chaper 12 describes in greater detail the influence of surface coatings on the slip resistance of bolted joints.

Table 5.1. Summary of Slip Coefficients

Type Steel	Treatment	Average	Standard Deviation	Number of Tests
A7, A36, A440	Clean mill scale	0.32	0.06	180
A7, A36, A440, Fe37, Fe52	Clean mill scale	0.33	0.07	327
A588	Clean mill scale	0.23	0.03	31
Fe37	Grit blasted	0.49	0.07	167
A36, Fe37, Fe52	Grit blasted	0.51	0.09	186
A514	Grit blasted	0.33	0.04	17
A36, Fe37	Grit blasted, exposed (short period)	0.53	0.06	51
A36, Fe37, Fe52	Grit blasted, exposed (short period)	0.54	0.06	83
A7, A36, A514, A572	Sand blasted	0.52	0.09	106
A36, Fe37	Hot-dip galvanized	0.18	0.04	27
A7, A36	Semipolished	0.28	0.04	12
A36	Vinyl wash	0.28	0.02	15
	Cold zinc painted	0.30	—	3
	Metallized	0.48	—	2
	Galvanized and sand blasted	0.34	—	1
	Sand blasted and treated with linseed oil (exposed)	0.26	0.01	3
	Red lead paint	0.06	—	6

5.1.7 Effect of Variation in Bolt Clamping Force

Besides the slip coefficient k_s, the initial bolt clamping force T_i is one of the major factors governing the slip load of a connection, as is apparent from Eq. 5.2. A variation in the initial clamping force directly affects the slip load of the connection. Experience has shown that the actual bolt tensions in a joint usually exceed the minimum tension required by specifications. This results from different tightening methods and variations in the mechanical properties of the bolts.

Bolts can be tightened by either the turn-of-nut method or with calibrated wrenches. The turn-of-nut method is primarily based on an elongation control, whereas the calibrated wrench method is based on controlling the applied torque. The two methods do not necessarily yield the same bolt tension, as illustrated in Fig. 5.5. Here the influence of the tightening method on the bolt tension achieved is shown for two bolt lots having different mechanical properties. When the calibrated wrench method is used, the bolt tension T_{iC} is about the same for both lots since the wrench is adjusted for each lot. However, if the turn-of-nut method is

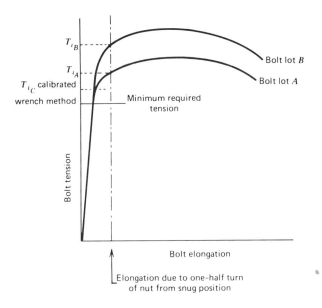

Fig. 5.5. Influences of tightening method on the bolt tension for different bolt lots.

employed, the average elongation of the bolts will be about the same for both lots. Consequently the bolt tensions T_{iA} and T_{iB} will differ, as illustrated in Fig. 5.5.

i. Turn-of-the-Nut Method. Figure 5.5 illustrates that the tensile strength of the bolt is a significant factor influencing the induced bolt tension when the turn-of-nut method is used. An increase in tensile strength leads to an increase in initial bolt tension in an installed bolt. An analysis of the data obtained from several bolt lots used in joints and calibration tests at Lehigh University indicates that the relationship between the tensile strength and initial bolt tension can be approximated by the straight line relationship given in Fig. 5.6. The tensile strength of a bolt was determined from static tension tests on representative samples. The induced bolt tension at one-half turn from the snug position can be derived from the measured average tensile force in bolts installed in joints or by torquing the bolts in an hydraulic calibrator. The data plotted in Fig. 5.6 show clearly that torquing a bolt one-half turn from the snug position in gripped material such as a joint leads to a higher tension stress than obtained by torquing the bolt one-half turn in an hydraulic calibrator. This is mainly due to the difference in stiffness of the gripped material as compared with the hydraulic calibrator.[4.1] Based on a least squares fit of all the data plotted in Fig. 5.6, the relationship between σ_i and σ_u was determined as

$$\sigma_i = 0.80\sigma_u \tag{5.4}$$

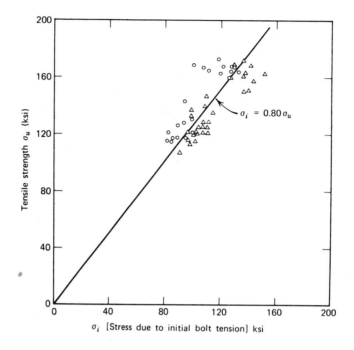

Fig. 5.6. T_i versus T_u in Lehigh tests. ○ Data from calibration tests; △ data from test joints.

Most of the data obtained from calibration tests in an hydraulic calibrator yield smaller bolt tensions compared with the data obtained from test joints (see Fig. 5.6). Hence, including the above data tends to yield a conservative estimate of the average bolt tension in a joint based on the average tensile strength of the bolts.

The actual bolt tension using the turn-of-nut method may exceed substantially the required minimum tension. This is illustrated in Fig. 5.7 where test data obtained from joints assembled with A325 bolts installed to one-half turn from snug are shown. The bolt tension on the horizontal axis is plotted as a percentage of the minimum required bolt tension. The average bolt tension in these joints was about 20% greater than the required minimum tension. In joints assembled with A490 bolts, installed to one-half turn from snug, an average bolt tension of 26% greater than the required minimum tension was observed. The bolts used in these tests were purposely ordered to minimum strength requirements of the applicable ASTM specification. Although the actual tensile strength of the bolts exceeded the required tensile strength (3% for A325 and 10% for the A490 bolts), it was less than the average tensile strength of production bolts.[4.5,4.6,5.12,5.25]

Since the average tensile strength of A325 bolts is

$$\sigma_{u\,\text{real}} = 1.183\sigma_{u\,\text{specified}}$$

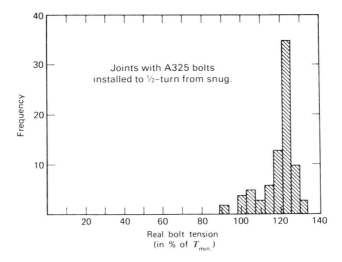

Fig. 5.7. Distribution of initial bolt force in test joints with A325 bolts installed to $\frac{1}{2}$ turn from snug. Number of tests, 81; average value, 120.2%; standard deviation, 9.1%.

and the average clamping force is about 80% of the actual tensile strength, it follows that the installed bolt tension σ_i is about equal to 0.95 $\sigma_{u\,\text{specified}}$. Present specifications require minimum bolt tension to equal or exceed 70% of the specified tensile strength. Hence, the average actual bolt tension will likely exceed the required minimum bolt tension by approximately 35% when the turn-of-nut method (one-half turn from snug) is used to install the bolts.

A similar analysis of A490 bolts installed to one-half turn from snug shows that the average initial bolt tension can be expected to exceed the minimum required bolt tension by approximately 26%.

Tests on short grip length high-strength bolts installed to one-third turn from snug yield similar values.[5.52] The results are shown in Fig. 5.8. The average bolt tension for short-grip A325 bolts was 26% greater than the required minimum tension. The results for short-grip A490 bolts show an even greater increase, but the number of data are very small. Other tests on short-grip A325 bolts installed to one-third turn from snug in coated joints indicated an average bolt tension 20% greater than that required.[5.53]

To characterize the frequency distribution of the ratio $T_i/T_{i\,\text{specified}}$, the standard deviations as well as the average values of the ratio are required. These have been estimated for both A325 and A490 bolts from test results. Data obtained at the University of Illinois, Lehigh University, and the University of Texas showed that the standard deviation of the ratio $T_i/T_{i\,\text{specified}}$ from average values was between 6 and 12% for A325 and A490 bolts. By assuming a normal distribution, the frequency distribution curve of the ratio $T_i/T_{i\,\text{specified}}$ can be defined. Figure 5.9 shows

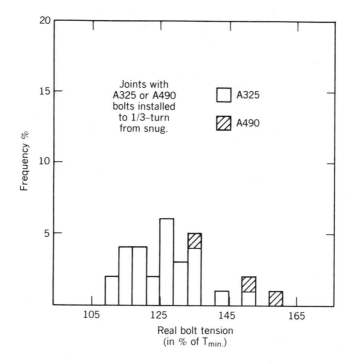

Fig. 5.8. Joints with A325 or A490 bolts installed to $\frac{1}{3}$ turn from snug.

these curves for A325 and A490 bolts. The figure shows that bolts installed by the turn-of-nut method will provide a bolt tension that exceeds the minimum required tension.

It was noted earlier that the average tensile strength of production A325 bolts exceeds the required tensile strength by approximately 18%. This was observed for bolt sizes up to 1-in. diameter. For A325 bolts greater than 1 in., the range of actual over specified minimum ultimate strength is even more favorable. The extra strength of bolts larger than 1 in. was not considered.

ii. Calibrated Wrench Method. A variation in mechanical properties of bolts does not affect the average installed bolt tension when the calibrated wrench is used. However, since this method is essentially one of torque control, factors such as friction between the nut and the bolt and between the nut and washer are of major importance. An analysis of 231 tests in which single bolts were subjected to a constant predetermined applied torque showed that the standard deviation of the recorded bolt tension equaled 9.4% of the recorded value.[4.13,5.35,5.36] It was observed that the variation of the average clamping force for a joint decreases depending on the number of bolts in the joint. For a joint having five bolts, the standard variation of the average bolt clamping force becomes 5.6% of the required mean value.

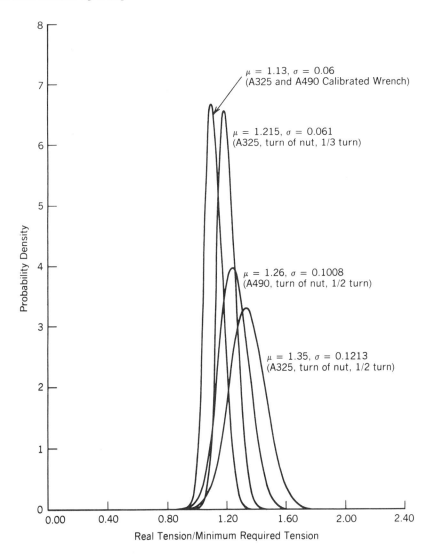

μ = 1.13, σ = 0.06
(A325 and A490 Calibrated Wrench)

μ = 1.215, σ = 0.061
(A325, turn of nut, 1/3 turn)

μ = 1.26, σ = 0.1008
(A490, turn of nut, 1/2 turn)

μ = 1.35, σ = 0.1213
(A325, turn of nut, 1/2 turn)

Fig. 5.9. Distribution of $T_i/T_{i\,\text{spec}}$ for different installation procedures.

Because variations in bolt tension do occur as a result of variations in thread mating, lubrication, and presence or absence of dirt particles in the threads, specifications usually require that the wrench be adjusted to stall at tensions 5 to 10% greater than the required preload.

Tests have indicated that installing a bolt in a joint leads to a higher bolt tension as compared with torquing the bolt in an hydraulic calibrator. This difference is about equal to 5.5%. Consequently the average clamping force in a five-bolt joint, with bolts installed by the calibrated wrench with a setting 7.5% greater than the required preload, is equal to

$$(0.7\sigma_u)(107.5)(1.055) = 0.796\sigma_u$$

or $1.13\sigma_{spec.\,min.}$. The standard deviation is equal to about 6%. The corresponding frequency distribution curve of the ratio $T_i/T_{i\,specified}$ for bolts installed by the calibrated wrench method is also shown in Fig. 5.9.

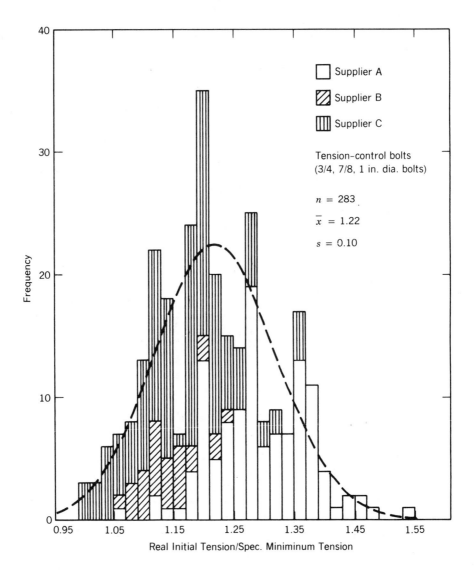

Fig. 5.10. Distribution of $T_i/T_{i\,spec}$ for A325 tension-control bolts.

iii.   Alternate Bolts.   The use of alternate bolts, load-indicating washers, or other nonstandard methods for introducing and monitoring the bolt preload will not necessarily lead to the same levels and distributions of preload as described here for bolts installed by the turn-of-nut method or by a calibrated wrench. Data are available for the tension-control type bolts described in Chapter 4. The results shown in Fig. 5.10 are for $\frac{3}{4}$-in., $\frac{7}{8}$-in., and 1-in. dia. A325 quality bolts obtained from three different suppliers. Distinct differences in the ratio of real initial tension to specified minimum tension can be seen, depending upon the supplier. Using all the test results, the mean value of the ratio is 1.22, about the same as that for A325 bolts installed to one-third turn from snug tight. The standard deviation from the mean is slightly larger for the tension-control bolts than for the normal A325 bolts.

5.1.8   Effect of Grip Length

The grip length of bolts does not have a noticeable influence on the behavior of friction-type joints. The only point of concern is the attainment of the desired clamping force. When the bolt length in the grip is greater than about eight times the diameter, one-half turn from the snug position may not provide the required preload. The greater bolt length requires an increased amount of deformation. To provide this increased bolt elongation, an additional increment of nut rotation is required. As was described in Section 4.1, the RCSC specification requires that the turn-of-nut be increased from one-half turn to two-thirds turn in order that at least the minimum bolt tension be reached in bolts with long grips.

Bolts with short grips are not likely to have less than the desired preload if installed by the turn-of-nut method. As noted in Section 4.1, they can, however, have a reduced rotational reserve if one-half turn is attempted. The RCSC specification prescribes one-third turn for bolts whose length is less than four diameters in order that the preload be developed and the rotational reserve maintained.

5.2   JOINT BEHAVIOR AFTER MAJOR SLIP

5.2.1   Introduction

When the frictional resistance of a joint is exceeded, a major slip occurs between the connected elements. Movement is stopped when the hole clearance is taken up and the bolts are in bearing. From this stage on, the load is mainly transferred by means of shear and bearing. This has led to the concept of a ''bearing-type'' joint. In bearing-type joints, the shear strength of the fasteners and the local bearing stresses in the plate around the fasteners are the critical parameters, not the bolt preload. As was noted in Subsection 4.2.2, the ultimate shear strength of high-strength bolts is not dependent upon the amount of preload in the bolts.

5.2.2   Behavior of Joints

The applied load in bearing-type joints may be transferred either by friction or by shear and bearing, depending on the magnitude of the load and the faying surface condition. In most joints a combination of both effects is likely to occur under normal service loads.

Initially, the load is transferred by friction forces at the ends of the joints. This is known from both elastic studies and from experimental investigations.[5.5] As the load is increased, the zone of friction extends toward the center of the joint as illustrated in Fig. 5.11. Eventually, the maximum frictional resistance is exceeded at the ends, and small displacements of contact points on the faying surfaces takes place. This is illustrated schematically in Fig. 5.11 as case 2. As load on the connection is increased, the slip zone proceeds inward from the ends toward the center of the joint. When the applied load exceeds the frictional resistance over the entire faying surface of the connection, large relative displacements occur. This movement, called major slip, theoretically may be equal to two hole clearances.

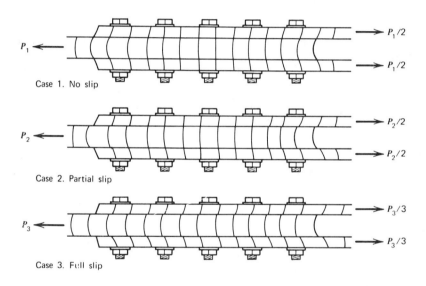

Schematic representation of three displacement conditions of a high-strength bolted joint.

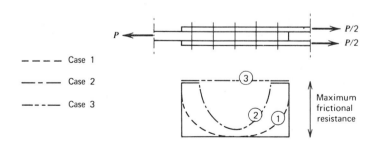

Fig. 5.11.   Distribution of friction forces for cases 1 to 3.

In practical terms it is observed to be much less than this. In laboratory tests it is usually about one-half a hole clearance, and measured values in the field are even smaller.[4.6,4.7,5.6,5.12,5.25,5.45]

When major slip occurs, only the end bolts may come into bearing against the main and splice plates. As the applied load is increased, the end bolts and holes deform further until the succeeding bolts come into bearing. This process continues until all of the bolts are in bearing, as illustrated for case 3 in Fig. 5.11.

Further application of load causes each bolt to deform in proportion to the force it transfers. The deformation of a bolt during this stage depends on the differences in plate elongations (main plate and lap plates) between any two adjacent transverse rows of bolts. Because the differential elongations are greater at the ends of the joint (e.g., the main plate may have yielded while the lap plates are still elastic), the end bolts are carrying greater loads than the interior bolts. A leveling out occurs if the bolts have good shear ductility, as is illustrated in Fig. 5.12a. Eventually the end pitches have such a large displacement and differential elongation that the end bolts fail in shear.

In short connections, with only a few fasteners in line, almost complete equalization of load is likely to take place before bolt failure occurs. Failure in this case appears as a simultaneous shearing of all the bolts.

In longer joints, the end fasteners will reach their critical shear deformation and fail before the full strength of each fastener can be achieved. The large shearing deformations of the end bolts and the greater elongation of the end holes is shown

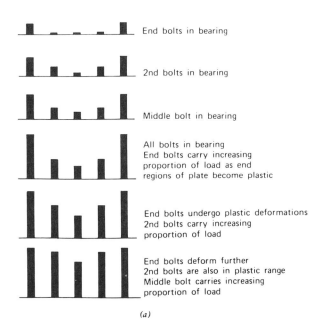

(a)

Fig. 5.12. (a) Bolt forces after major slip; (b) sawed end sections of bolted joint.

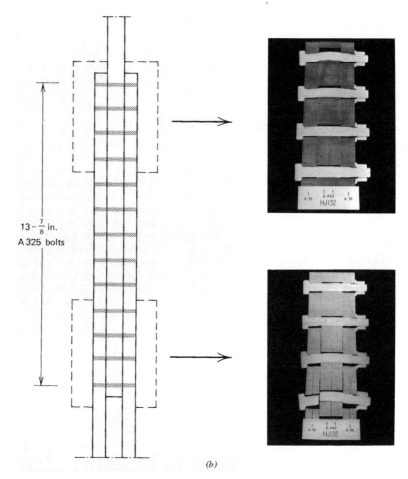

13 - $\frac{7}{8}$ in.
A 325 bolts

(b)

Fig. 5.12.   (*Continued*)

in Fig. 5.12*b*. The remaining bolts are usually not capable of taking much additional load without incurring failure themselves in a sequential fashion. The sequential failure of fasteners in long connections is called "unbuttoning." This phenomenon is predicted by theoretical analysis and has been witnessed in tests of long bolted and riveted joints.[4.6,4.7,5.6,5.12,5.21,5.25]

Figure 5.13 shows load versus formation curves for two A7 steel ($\sigma_y = 33$ ksi) joints connected with A325 bolts. Figure 5.13*a* is the test curve of a joint with semipolished faying surfaces. A gradual slip occurred as load was applied. The second joint had clean mill scale surfaces and exhibited a sudden slip, as illustrated in Fig. 5.13*b*.

High-strength bolts are usually placed in holes that are nominally $\frac{1}{16}$ in. larger than the bolt diameter. Therefore, the maximum slip that can occur in a joint is

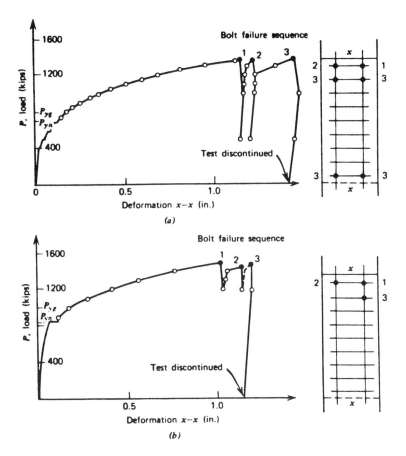

Fig. 5.13. Typical load versus deformation curves for different surface conditions. (*a*) Semipolished surfaces; (*b*) clean mill scale surfaces.

equal to $\frac{1}{8}$ in. However, field practice has shown that joint movements are rarely as large as $\frac{1}{8}$ in. and average less than $\frac{1}{32}$ in.[5.45] In many situations the joint will not slip at all under live loads because the joint is often in bearing by the time the bolts are tightened. This might be due to small misalignments inherent to the fabrication process. In addition, slip may have occurred under the dead load before the bolts in the joint were tightened. Generally, slips under live loads are so small that they seldom have a serious effect on the structure.

Bolt preload is obtained by the introduction of a relatively small axial elongation of the bolt as the nut is turned. As a bolt loaded in shear approaches its ultimate load, the relatively large shearing deformations that have occurred have the effect of releasing the axial elongation that was used to obtain bolt preload. Thus, there is practically no preload in the bolt at time of failure by shear rupture.[4.6,4.7] As a consequence, there is also negligible frictional resistance at the time the ultimate load is reached.

5.2.3 Joint Stiffness

The stiffness of a bearing-type joint is equal to the stiffness of similar slip-resistant joints until slip occurs. Slip of the connection brings one or more bolts into bearing and results in motion of the lap plates with respect to the main plates. The stiffness of the joint, characterized by the slope of the load versus elongation curve, is not affected by slip. This is illustrated in Fig. 5.14. Only yielding of the gross and net sections caused a significant change in the slope of the load versus elongation curve.

The load versus deformation curves shown in Figs. 5.13b and 5.14 show a distinct slip. In most situations the slips are so small that they have no significant effect on the structure. The joint stiffness of a bearing-type joint is about the same as the stiffness of a similar slip-resistant joint if the joint is erected in bearing.

5.2.4 Surface Preparation and Treatment

The level of slip resistance does not influence the ultimate strength of a shear splice (see Fig. 5.13). Consequently, the surface condition of the connected plates is not critical except for slip-critical joints. Hence, paint, galvanization, or other surface conditions that may result in a low slip coefficient do not influence the ultimate strength of bolted joints.

The slip resistance is an important factor influencing the joint behavior under

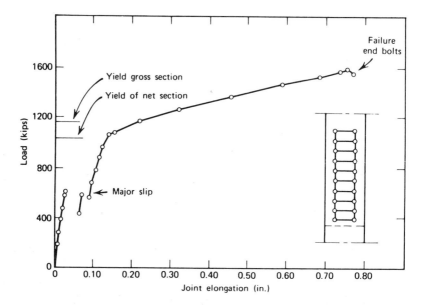

Fig. 5.14. Typical load versus deformation curve of high-strength bolted joint.

repeated loadings. Depending on the ratio between the slip resistance and applied load, failure may occur through either the net or gross section area. A more detailed discussion on this is given in Section 5.3.

5.2.5 Load Partition and Ultimate Strength

Theoretical studies of mechanically fastened joints have been made since the beginning of this century. A linear, elastic relationship between load and deformation was assumed in early studies. However, since the early 1960s, mathematical models that establish the relationships between deformation and load throughout the elastic and inelastic range for component parts of joints have been developed.[5.21] The method of analysis is summarized briefly in this section for a double shear symmetrical butt joint. For purposes of analysis, the joint is divided into gage strips, and it is assumed that all gage strips are identical in behavior. Test results have indicated that this is a reasonable approximation.

The theoretical solution of the load partition at ultimate load is based on the following major assumptions: (1) the fasteners transmit all the applied load by shear and bearing once major slip has occurred, and (2) the frictional forces may be neglected in the region for which the solution is intended, that is, the region between major slip and ultimate load.

The solution is obtained by formulating the following two basic conditions: (1) satisfying the condition of equilibrium, and (2) assuring that continuity will be maintained throughout the joint length for all load levels. These conditions, coupled with initial value considerations such as the ultimate strength of the plate and the ultimate strength and deformation capacity of the critical fastener, yield the solution.

The equilibrium conditions can be visualized with the aid of Fig. 5.15a. The load per gage strip in the main plate between bolts i and $i + 1$ is equal to the total load on this strip, P_G, minus the sum of the loads on all bolts, ΣR_i, preceding the part of the joint considered, that is, between i and $i + 1$:

$$P_{i,i+1} = P_G - \sum_{i=1}^{i} R_i \qquad (5.5)$$

The load per gage in the lap plates between bolts i and $i + 1$ is equal to the sum of the loads transmitted to the lap plate by all the bolts preceding the part of the joint considered. Hence

$$Q_{i,i+1} = \sum_{i=1}^{i} R_i \qquad (5.6)$$

The compatibility equations can be formulated by considering the deformations illustrated in Fig. 5.15b. As a result of the applied load, the main plate will have elongated so that the distance between the main plate holes is $p + e_{i,i+1}$. The lap plates will also have elongated, and the distance between the lap plate holes is

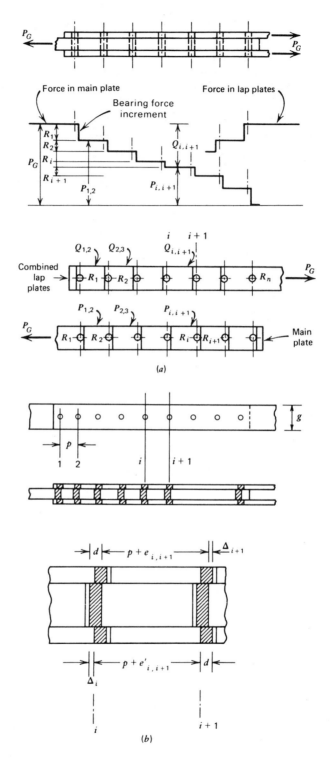

Fig. 5.15. Idealized load transfer diagrams and deformations in bolts and plates. (*a*) Load transfer; (*b*) deformations in bolts and plates.

$p + e'_{i,i+1}$. The bolts will have undergone deformations Δ_i, which include the effects of shear, bending, and bearing of the fastener and the localized effect of bearing on the plates. It is assumed that the deformations of the fastener, Δ_i, are the same whether considered at the hole edge (fastener surface) or the center line of the fastener. A further, detailed, analysis of the parameters included in Δ_i and Δ_{i+1} is given in Refs. 5.21 and 5.22.

The compatibility condition between points i and $i + 1$ yields

$$\Delta_i + e'_{i,i+1} = \Delta_{i+1} + e_{i,i+1} \tag{5.7}$$

If the plate elongations are expressed as functions of load in the segments of the joint between fasteners, and the fastener deformations as functions of the fastener loads, Eq. 5.7 can be written as

$$f(R_i) + \Psi(Q_{i,i+1}) = f(R_{i+1}) + \Phi(P_{i,i+1}) \tag{5.8}$$

in which $f(R_i)$ and $f(R_{i+1})$ represent the bolt deformations, $\Phi(P_{i,i+1})$ the main plate elongation, and $\Psi(Q_{i,i+1})$ the lap plate elongation.

Equation 5.8 can be written for each section of the joint, giving $n - 1$ simultaneous equations. These, with the equation of equilibrium,

$$P_G - \sum_{i=1}^{n} R_i = 0 \tag{5.9}$$

may be solved to give the loads acting on the fastener when the relationships between the load and elongation for the various components are known.[5.21,5.22] With this information, the total load acting on the joint can be found for a given end fastener deformation. The ultimate strength, the load at failure, can be found by setting the deformation of the end fastener equal to its ultimate deformation.

The solution of the equilibrium and compatibility equations is lengthy and laborious, especially for long joints with many fasteners. Obviously, such solutions are not practical for design purposes. However, the theoretical solution for the ultimate strength and load partition has been accomplished by computer studies and verified by comparing the theoretical results with the results of tests of large steel joints with yield strengths ranging from 33 to 100 ksi.[4.6,5.6,5.12] In all cases, the theory and test results were in good agreement. Fig. 5.16 shows the experimental and theoretical load versus deformation curve for a bolted joint with two lines of ten $\frac{7}{8}$-in. diameter A325 bolts per line. The yield stress of the plate material was about 44 ksi, and the ratio of the net section area to the gross section area, denoted as the A_n/A_g ratio, was 1.10 for this particular joint. The theoretical loads carried by each fastener at two stages of loading are shown in Fig. 5.17. The end fasteners are obviously the critical ones.

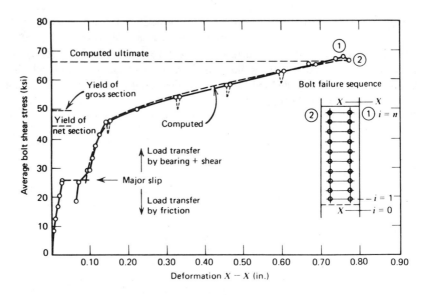

Fig. 5.16. Comparison of theoretical and experimental results.

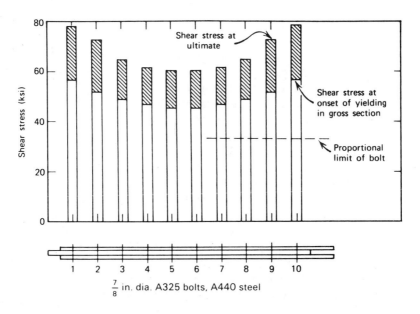

Fig. 5.17. Load partition in joint with 10 fasteners in a line.

98

5.2.6 Effect of Joint Geometry

By means of the theoretical solution summarized in Subsection 5.2.5, it is possible to study the effect of material and geometrical parameters that govern the joint behavior. In this article the significance of a number of parameters such as the joint length, the pitch, the relative proportions between the net tensile area of the plate and the total bolt shear area (A_n/A_s ratio), the type of connected material, the A_n/A_g ratio, and the fastener pattern are examined briefly. A more detailed analysis of these parameters is presented in Refs. 5.21 and 5.23. All the hypothetical studies are based on minimum strength plate and fasteners and provide a lower bound to the joint strength.

i. Effect of Joint Length.   Theoretical as well as experimental studies have shown that the joint length is an important parameter that influences the ultimate strength of the joint. Depending on factors such as type of plate material and fastener deformation capacity, a simultaneous shearing of all the bolts or a sequential failure (unbuttoning) of all the bolts may occur, depending on the joint length.

For a given number of fasteners, the joint length is a function of the fastener spacing (pitch). A constant pitch of $3\frac{1}{2}$ in. and a ratio of bolt shear area to net tensile area equal to 1.10 were used in theoretical studies to illustrate the effect of joint length. The joint material has a yield strength of 36 ksi and it is fastened by $\frac{7}{8}$-in. dia. A325 bolts. If the design stress of the plate material is taken as 24 ksi, then an A_n/A_g ratio of 1.10 yields an average shear stress of about 22 ksi for the fasteners.

The results of the theoretical studies are summarized in Fig. 5.18, where the average fastener shear at ultimate load is plotted as a function of the joint length. The longer joints showed a significant decrease in average bolt shear strength as compared with the shear strength of a single fastener. Short or "compact" joints were affected to a negligible extent. Joints up to 10 in. in length provided about the same average shear strength as a single fastener. As the number of fasteners was increased, Fig. 5.18 indicates that a decrease in the average strength occurred at a decreasing rate.

The reason for the decrease in shear strength with increased length of the joint is illustrated in Figs. 5.19, 5.20, and 5.21. The computed shear stresses in each bolt at two different loading stages are shown for joints having 4, 10, and 20 fasteners in a line. The two stages are (1) onset of yielding in the gross section of the plate (designated by the end of the open portion of the bar), and (2) bolt stress at ultimate load (designated by the top of the shaded portion). Figure 5.19 shows that almost complete redistribution of bolt forces has taken place in the four-bolt joint, since all fasteners are carrying about the same load at ultimate. As joint length is increased, Figs. 5.20 and 5.21 show that the fasteners near the center of the joint carry only about half the forces carried by the end fasteners. Consequently, the average shear stress on the fastener is significantly reduced.

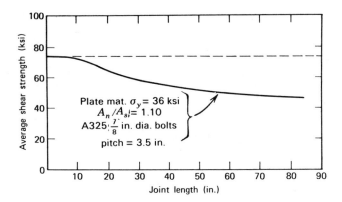

Fig. 5.18. Effect of joint length on ultimate strength.

Theoretical investigations to determine the influence of joint length on the load distribution in joints of steel with a yield stress other than 36 ksi have been made.[5.23-5.25] Steels with a yield stress ranging from 36 to 100 ksi, as well as hybrid steel joints, were examined, and the results indicated a load distribution similar to the one described previously for a 36 ksi yield stress plate material.

ii. Effect of Pitch. The pitch is the distance between centers of adjacent fasteners along the line of principal stress. To determine the effect of the fastener pitch, analytical studies were made for joints with different fastener spacings, bolt grades, and connected material.[5.6,5.21] The results of an analysis of a 36 ksi yield stress plate material connected by $\frac{7}{8}$-in. diameter A325 bolts are summarized in

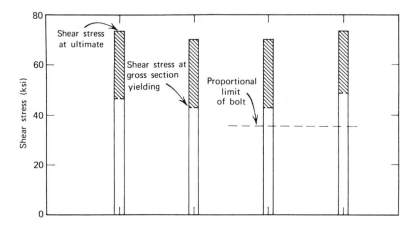

Fig. 5.19. Load partition in joint with four fasteners in line. Plate material $\sigma_y = 36$ ksi. Fastened by $\frac{7}{8}$-in. dia. A325 bolts.

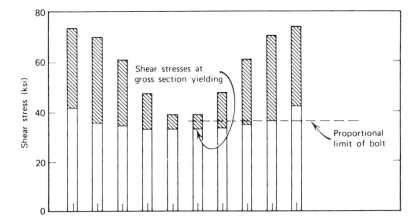

Fig. 5.20. Load partition in joint with 10 fasteners in line. Plate material $\sigma_y = 36$ ksi. Fastened by $\frac{7}{8}$-in. dia. A325 bolts.

Fig. 5.22. Three different fastener spacings, three, four, and seven times the bolt diameter, were examined. The curves indicate that the change in shear strength with length is not greatly influenced by the pitch of the fasteners. If a joint with a given number of fasteners in a line is shortened by reducing the pitch between bolts, equal or greater strength results from the decrease in length. These studies have shown that pitch length, per se, is not an important variable. For a given A_n/A_s ratio, the shear strength is controlled by total joint length rather than by pitch length.

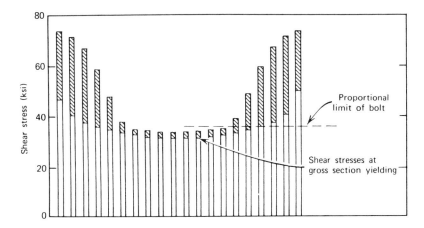

Fig. 5.21. Load partition in joint with 20 fasteners in line. Plate material $\sigma_y = 36$ ksi. Fastened by $\frac{7}{8}$-in. dia. A325 bolts.

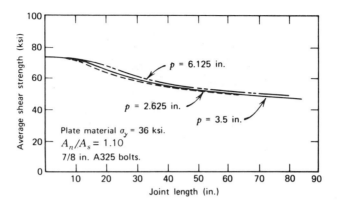

Fig. 5.22. Effect of pitch on the ultimate strength of steel joints. Plate material $\sigma_y = 36$ ksi. $A_n/A_s =$ 1.10, $\frac{7}{8}$-in. dia. A325 bolts.

iii. Effect of Variation in Relative Proportions of Shear and Tensile Areas.

There are two possible modes of fracture in a bearing-type connection subjected to a tensile load. If the differential plate strains near the ends of a joint are high as compared with those in the central portion, the shear failure of a single end fastener can occur. The resulting distribution of load from the failed connector to those remaining usually causes a sequential failure or unbuttoning, and little, if any, additional strength is available. When the tensile capacity of the plate at its net section is less than the shear capacity of the fasteners, failure will obviously occur by fracture of the plate.

Establishing the plate failure-fastener failure boundary line cannot be done directly, since joint length and the ratio of shear to net area both influence the shear strength. When the bolt shear strength and the plate capacity converge, a point on the boundary is determined. This process can be repeated for various joint lengths until the complete curve has been defined, as shown in Fig. 5.23. For comparative purposes, curves for steels with a yield stress of 50 and 100 ksi are shown.

It has been theoretically predicted and experimentally verified that, as the A_n/A_s ratio for a joint is increased for any given joint length, the average shear strength also increases.[5.23] Figure 5.24 summarizes the results of analytical studies on joints of a plate material having a 36 ksi yield stress and fastened with A325 bolts. An increase in the A_n/A_s ratio corresponds to an increase in the net tensile area. The ideal case of equal load distribution among fasteners occurs when $A_n/A_s = \infty$. This represents a perfectly rigid joint. For any lesser value of A_n/A_s, the fasteners carry unequal load, depending on the joint length. Figure 5.25 shows the effect of a variation in the A_n/A_s ratio for joints fastened by A490 bolts. A yield stress of 100 ksi was assumed for the plate material.

Both Figs. 5.24 and 5.25 illustrate that, with an increase in the net plate area, the average shear strength of the fasteners for the longer joints is greater. For

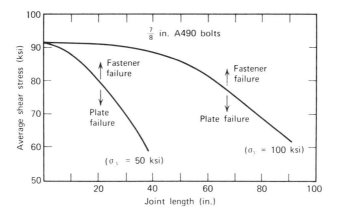

Fig. 5.23. Failure mode boundary (τ vs. L).

shorter joints, plate failure may occur before bolt failure. Only an increase in joint length can cause bolt failure.

This examination has illustrated that it is not possible to maintain a uniform condition for both bolts and plates. When joints are short, the usual plate geometry will cause plate failure to occur. As joint length is increased, a balanced condition

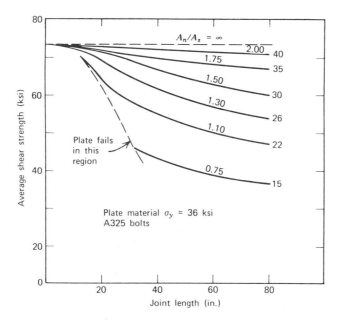

Fig. 5.24. Effect of variation of A_n/A_s ratio: structural carbon steel fastened by A325 bolts.

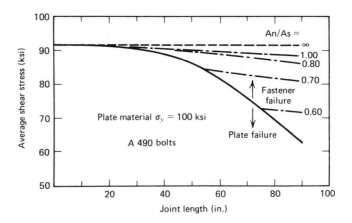

Fig. 5.25. Effect of variation of A_n/A_s ratio: quenched and tempered alloy steel fastened by A490 bolts.

can occur for a specified length. For longer joints, bolt failure will be the governing mode. For design, the achievement of a proper balance between these failure conditions is required.

iv.   Effect of Variation in Gage Width and A_n/A_g Ratio.   In evaluating the performance of any structure, it is usually considered desirable for the system to have the capacity for distortion or geometrical adjustment before failure by fracture. In an axially loaded structure, this means that, if at all possible, the connections should permit yielding to occur in the gross cross-section of the member before the joint fails through the net section.[5.23] This requirement is satisfied if

$$\frac{A_n}{A_g} \geq \frac{\sigma_y}{\sigma_u} \tag{5.10}$$

It is apparent that, depending on the type of steel, Eq. 5.10 leads to different minimum A_n/A_g ratios. Based on the specified minimum yield and tensile strengths for the type of steel, the A_n/A_g ratio has to equal or exceed 0.60 or 0.69 for structural carbon steel and high-strength steel, respectively, to achieve yielding of the gross section before failure of the net section occurs. For joints made of quenched and tempered alloy steel, the minimum A_n/A_g ratio is equal to 0.87.

These A_n/A_g requirements are satisfied in most structures of carbon or high-strength steel. However, it has been shown that for A514 steel (yield stress 90 to 100 ksi) tension members, current practice commonly will lead to the situation wherein the member will fail through the net section before yielding is reached in the gross section, unless special provisions such as upset ends or other changes in cross section are made to ensure yielding of the gross section before the net section fails.[5.23,5.38] If yielding in the gross section cannot be achieved, a greater margin against ultimate is needed.

The A_n/A_g ratio depends on factors such as the gauge width of the joint and the hole diameter. For a constant hole diameter, an increase in the gauge width g increases the A_n/A_g ratio; therefore, gross section yielding is more likely to occur before failure of the net section. An increase in gauge width also tends to decrease slightly the tensile strength of the net section. However, this is not critical, since gross section yielding of the member can be expected.

When a ductile metal bar is loaded and the resulting nominal stresses are plotted as a function of the strain, the characteristic relationship shown in Fig. 5.26 is observed. If a similar test is conducted on a tensile specimen with holes, the stress-strain relationship is modified, as also illustrated in Fig. 5.26. For the so-called plate calibration coupon, the average strain between the two holes has been used. The ultimate strength of perforated plates at the net section is higher than the coupon ultimate strength. This results because free lateral contraction cannot develop; the increase is attributed to the "reinforcement" or bi-axial stress effect created by the holes.[5.46] As the gauge is increased, this effect is less noticeable. Figure. 5.27 illustrates this behavior for different steels. The ratio $\sigma_u/\sigma_{u\,\mathrm{coup}}$ is plotted as a function of both the $g/(g-d)$ ratio and its reciprocal, A_n/A_g. From this plot it can be concluded that a decrease in the $g/(g-d)$ ratio (hence an increase in the A_n/A_g ratio) tends to decrease the ultimate strength of the net section.

v. Effect of Type of Connected Material. The yield stress of the connected material is known to influence the ultimate strength of a joint. For a given load and resulting number of bolts, the bolt shear area is constant, whereas the net and gross areas will change depending on the type of steel used in the joint. For a given load, an increase in yield stress of the plate material results in a decrease in the plate area. Since different plate areas are required, the A_n/A_s ratio of a joint is affected. The influence of an increase in yield stress of the plate material on the ultimate joint strength is illustrated in Fig. 5.28. The allowable shear stress on A325 bolts is assumed to be 30 ksi, and the allowable tensile stress for the plates is taken as 22 or 30 ksi for steel with a yield stress of 36 or 50 ksi, respectively. Employing the higher strength steel reduces the net area of the joint by a factor 22/30. Since the bolt shear area remains constant, the A_n/A_s ratio is reduced by the same factor and for this particular joint becomes equal to 1.0.

It is apparent from the comparison made in Fig. 5.28 that an increase in steel strength slightly decreases the joint strength because of the decrease in the A_n/A_s ratio. The difference is not large, however, and the lower bound provided by the higher strength steels can be used to develop design criteria.

vi. Fastener Pattern and Net Section Strength. Designing a tension member requires the selection of a section with sufficient net area to carry the working load (allowable stress design) or the factored load (load factor design) without exceeding the prescribed permissible stresses. Besides meeting this requirement,

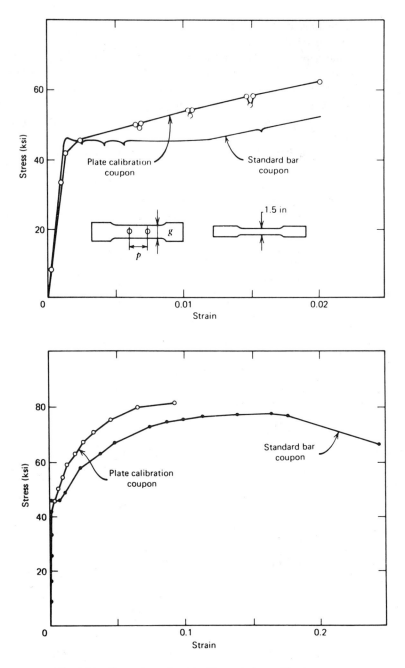

Fig. 5.26. Comparison of standard bar and plate calibration coupon.

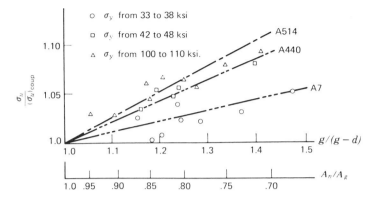

Fig. 5.27. Effect of A_n/A_s ratio on ultimate strength of tension specimen.

which is based on strength of the connection, it is usually considered desirable that tension members yield on the gross section before failure occurs at the net section. The A_n/A_g ratio reflects this requirement.

One of the parameters that influences the net area is the hole pattern. Often a simple rectangular pattern of fasteners is all that is necessary. However, in many

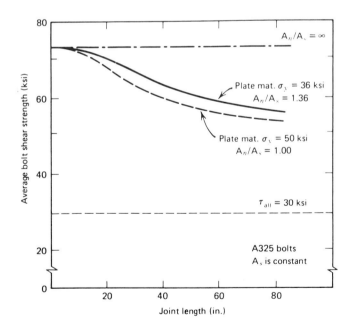

Fig. 5.28. Effect of type of connected material.

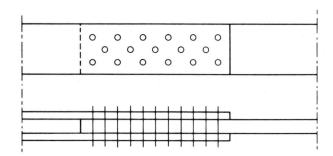

Fig. 5.29. Staggered fastener pattern.

situations a staggered hole pattern, as shown in Fig. 5.29, is required to satisfy the A_n/A_g requirement and increase the joint efficiency. For the rectangular pattern shown in Fig. 5.30a, failure is likely to occur through section A-A. The reduction in area will be directly related to the diameter of the two holes. If the critical cross-section is analogous to case c, failure will occur at section C-C, and the reduction in area will be caused by only one hole. It is more likely that the actual failure will be bounded by these two conditions. Case b represents this intermediate reduction in joint capacity. The area to be deducted is a function of the stagger s

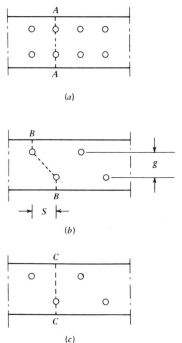

Fig. 5.30. Possible failure paths for different hole patterns.

and the gauge g. The following function was developed by Cochrane in 1922 and is widely used for the design of tension connections:

$$A_n = t\left(W_g - nd + \Sigma \frac{s^2}{4g}\right) \tag{5.11}$$

where W_g describes the gross width of the member.[5.26] With this equation the net section of a flat plate-type joint with a staggered hole pattern can be evaluated with reasonable accuracy.[5.27-5.29]

If a tension member is to yield on the gross section before failure occurs at the net section, the following equation must be satisfied

$$A_n \sigma_u \phi \geq A_g \sigma_y \tag{5.12}$$

where σ_u and σ_y represent the tensile strength of the net section and the yield stress of the material at the gross section; ϕ is a reduction factor to ensure that yielding of the gross section develops before the tensile capacity of the net section is reached. For design purposes it is convenient to express Eq. 5.12 as

$$\frac{A_n}{A_g} \geq \frac{\sigma_y}{\phi \sigma_u} \tag{5.13}$$

It is shown in Fig. 5.27 that the tensile strength of a plate with holes depends on the A_n/A_g ratio as well as on the type of steel; for the practical range of A_n/A_g ratios, the tensile strength σ_u of the net section will exceed the plate coupon tensile strength by about 7 or 8%. Consequently, using the coupon strength σ_u in Eq. 5.13 yields a conservative A_n/A_g ratio for a rectangular fastener pattern. If a staggered hole pattern is used, the net section is determined from Eq. 5.11. Since Eq. 5.11 is based on test results, the constraining effect of the hole pattern is automatically included.

To ensure that yielding on the gross section does occur before failure of the net section and also to provide a minimum factor of safety against a net section tensile failure, a reduction factor ϕ is required; ϕ also prevents yielding of the net section under working loads.

This examination indicates that the net section need not to be considered as the critical design section if Eq. 5.13 is satisfied. When Eq. 5.13 cannot be satisfied, the design must ensure a satisfactory margin against failure of the net section. Most of the quenched and tempered alloy steel joints do not meet the requirements of Eq. 5.13 and are to be designed on the basis of adequate net section strength.

5.2.7 Type of Fastener

Often situations arise where the type of fastener may be variable; that is, either A325 or A490 bolts can be used. A change in bolt type corresponds to a change in the A_n/A_s ratio when the net area of the joint is maintained, since the required number of bolts must change. The effect of changing the bolt type is illustrated in Fig. 5.31. Figure 5.31*a* corresponds to allowable bolt shear stresses of 22 and 32 ksi for A325 and A490 bolts, respectively, Figure 5.31*b* corresponds to allowable shear stresses of 30 and 40 ksi for the same bolts. (The bolt shear stresses used in Fig. 5.31*a* are those that were permitted for working stress design in the 1973 RCRBSJ specification. The values used in Fig. 5.31*b* are those permitted in the 1985 RCSC specification for the same case.) The yield stress of the plate material

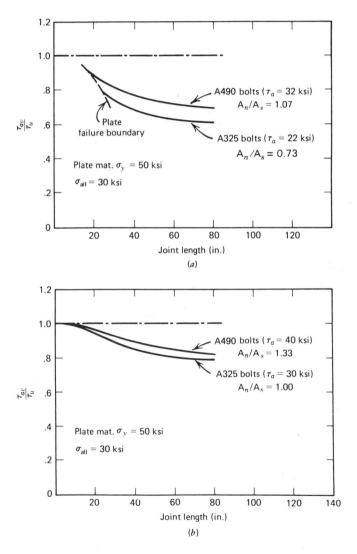

Fig. 5.31. Effect of type of fastener. (*a*) Behavior for 1973 design stresses. (*b*) Behavior for 1985 design stresses.

was assumed to be equal to 50 ksi; this resulted in an allowable stress of 30 ksi for the plate material. By employing A490 instead of A325 bolts, the bolt shear area is significantly reduced, and consequently the A_n/A_s ratio is increased. The increase in the A_n/A_s ratio provides a more favorable conditions for the longer joints. The increase in efficiency is not as significant for shorter joints.

Besides the increase in the A_n/A_s ratio, a change from A325 to A490 bolts also reduces the joint length for a given design load. This often provides a more favorable joint condition.

5.2.8. Effect of Grip Length

For joints with up to 6 in. of gripped material, test results are in close agreement with the analytical solution. Joints with larger grips and longer bolts tend to give higher ultimate loads than predicted.[5.25]

A qualitative explanation for this observed behavior can be developed from the sheared bolts shown in Fig. 5.32. Shear tests of single bolts yield shear planes at almost 90° to the bolt axis when rigid plate elements are used, whereas the bolt from a joint with a large grip fails along an inclined shear plane. In joints fastened with long bolts, the individual plates adjust to the loads they carry, and the bolts assume the curved shape shown in Fig. 5.33. This results in an increased shearing area and increases the ultimate load and deformation capacity of the bolt. Hence, the end fastener in a joint with long bolts deforms more than expected and permits the interior bolts to carry more load.

The extent to which a bolt bends is affected by the slippage of the plates with respect to one another. Furthermore, the number of plies within the grip length of the fasteners is an important factor in developing fastener bending. For joints with high A_n/A_s ratios, the bending is more pronounced in more bolts, as illustrated in

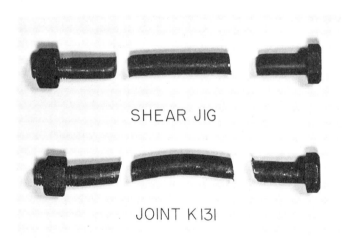

Fig. 5.32. Comparison of sheared bolts.

Fig. 5.33. Sawed sections of joints showing bolt bending.

Fig. 5.33. This results in an increased joint strength if failure occurs in the fasteners.

5.2.9 Bearing Stresses and End Distance

Failure of a bolted or riveted joint occurs if the applied load exceeds (1) the tensile capacity of the critical net section, (2) the shear capacity of the fasteners, or (3) the bearing strength of the material. The net section strength as well as the fastener shear strength were examined earlier. This section deals specifically with failures related to high bearing stresses on the fastener and the plate material.

After major slip has occurred in a connection, one or more fasteners are in bearing against the side of the hole. A bearing stress is developed in the material adjacent to the hole and in the fastener, as shown in Fig. 5.34a. Initially, this stress is concentrated at the point of contact. An increase in load causes yielding and the embedment of the bolt on a larger area of contact, and this results in the more uniform stress distribution indicated in Fig. 5.34b. Although the actual bearing stress distribution is not known, a uniform stress distribution can be assumed as indicated in Fig. 5.34c. The nominal bearing stress can be expressed as

$$\sigma_b = \frac{P}{dt} \tag{5.14}$$

where P denotes the load transmitted by the fastener, t the plate thickness, and d the nominal bolt diameter. Although the fastener itself is subjected to the same

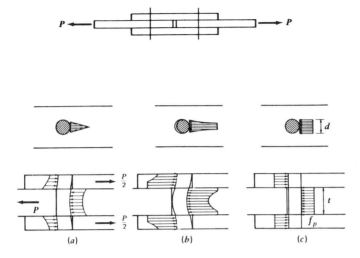

Fig. 5.34. Bearing stresses. (*a*) Elastic. (*b*) Elastic-plastic. (*c*) Nominal.

magnitude of compressive forces as those acting on the side of the hole, tests have always shown that the fastener is not critical.[5.31,5.32,5.39,5.40]

The actual failure mode in bearing depends on such geometrical factors as the end distance, the bolt diameter, and the thickness of the connected plate material. Either the fastener splits out through the end of the plate because of insufficient end distance, as illustrated in Fig. 5.35*a*, or excessive deformations are developed in the material adjacent to the fastener hole, as indicated in Fig. 5.35*b*. Often a combination of these failure modes will occur.

The end distance required to prevent the plate from splitting out can be estimated by equating the maximum load transmitted by the end bolt to the force that corresponds to shear failure in the plate material along the dashed lines in Fig. 5.35*c*. The maximum shear capacity of a single bolt is equal to

$$P_S^b = mA_b\tau_u^b \tag{5.15}$$

where m is equal to the number of shear planes. The load on the fastener is also represented as

$$P_S^b = td\sigma_b \tag{5.16}$$

A lower bound to the shear resistance developed along the dashed lines (Fig. 5.35*c*) can be expressed as

$$P_S^P = (2t)\left(L - \frac{d}{2}\right)(\tau_u^P) \tag{5.17}$$

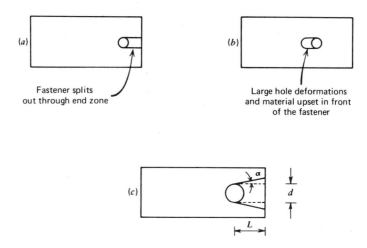

Fig. 5.35. Failure modes. (*a*) Fastener splits out through end zone. (*b*) Large hole deformations and material upset in front of the fastener.

where τ_u^P represents the shear strength (in kilopounds per square inch) of the plate material. For most commonly used steels, the shear strength is about 70% of the tensile strength. Hence Eq. 5.17 can be transformed into

$$P_S^P = (2t)\left(L - \frac{d}{2}\right)(0.7\sigma_u^P) \tag{5.18}$$

where σ_u^P represents the tensile strength of the plate material and L the end distance of the fastener. A lower bound to the L/d ratio that will prevent the fastener from splitting out of the plate material is obtained from Eqs. 5.16 and 5.18, namely

$$\frac{L}{d} \geq 0.5 + 0.715\frac{\sigma_b}{\sigma_u^P} \tag{5.19}$$

This equation relates the bearing ratio σ_b/σ_u^P to the end distance represented by the L/d ratio.

Figure 5.36 shows the analytical solution provided by Eq. 5.19 as compared with test results. Included in the tests are one, two, and three bolt specimens and one rivet or three rivet specimens.[5.31,5.32,5.39,5.40,5.53] The bolted specimens were untightened (finger tight) or tightened, as noted. In most cases the outside plates of these butt joints were critical, but in a few cases the inner ply was critical. It should be noted that the bearing ratio has been plotted as a function of L/d for all cases except for the two-bolt specimens (nontightened) that failed on the enclosed ply. In this situation, the parameter L was replaced by $(s - 0.5d)$, where s is the bolt pitch.

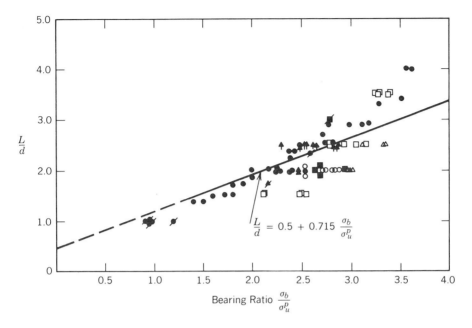

Fig. 5.36. Influence of type of specimen on the bearing. ● One-rivet (or bolt) specimen (non-tightened); ○ one-bolt specimen (tightened): ▲ two-bolt specimen (nontightened); △ two-bolt specimen (tightened); ■ three-rivet (or bolt) specimen; □ three-bolt specimen (tightened). Note: All specimens except ⤴⤴▲■ were critical on enclosed ply; the ⤴⤴▲■ specimens had outside plates critical.

There is reasonably good agreement between the prediction provided by Eq. 5.19 and the test results over most of the range. However, the analytical solution and the test results tend to diverge at the larger L/d ratios. This is expected because an increasing L/d ratio will gradually change the failure mode. For high L/d ratios, failure will not occur by shearing out the plate material in the end zone, as was assumed in the analytical solution. Failure will occur by the material piling up as indicated in Fig. 5.35*b*.

A closer examination of the data represented in Fig. 5.36 will show individual differences depending upon the type of test specimen. One bolt specimens (one bolt on either side of the splice) are more critical than two bolt or three bolt specimens, for example. It is also apparent that providing a clamping force in the bolt leads to an increase in the ultimate bearing ratio. This indicates that the load is partially transmitted by frictional resistance on the faying surfaces. Consequently, the real bearing stress is less than the "ultimate" bearing stress computed on the basis of the total applied load. Most the data summarized in Fig. 5.36 were obtained from tests on symmetric butt joints. Failure always occurred in the main plate in these. If the lap plates are relatively thin compared with the main plate, then failure may occur in the lap plates, however. Test results have indicated that in these situations bearing failures are influenced by "catenary action," which causes bending in the lap plates.[5.31,5.40] The thin lap plates bend outwards and decrease the ultimate bearing strength of the connection. A series of tests in which very thin lap plates were used is contained within the results presented in Fig. 5.36. In this program, $\frac{1}{4}$-in. thick lap plates were used in conjunction with a 2-in. thick main plate. The end distance used was 9 in., which exceeds by a significant margin the maximum permissible end distance permitted by most specifications. Nonetheless, the results are still in reasonable agreement with other test results where the enclosed plies are critical.

5.3 JOINT BEHAVIOR UNDER REPEATED LOADING

5.3.1 Basic Failure Modes

The behavior of a bolted connection under repeated loading is directly influenced by the type of load transfer in the connection. The applied load can be transferred either by friction on contact surfaces, by shear and bearing of the bolts, or by both, depending on the direction of the applied load, the magnitude of the clamping force, the condition of the faying surfaces, and the possible occurrence of major slip. Tests have shown that each load transfer mechanism develops its own characteristic failure pattern under repeated loadings.[5.18] These characteristic conditions are best explained and illustrated by examining the stress distribution throughout the joint.

Figure 5.37 shows schematically an idealized lap joint subjected to a cyclic, in-plane force. Assuming that no major slip occurs, hence that the external load is

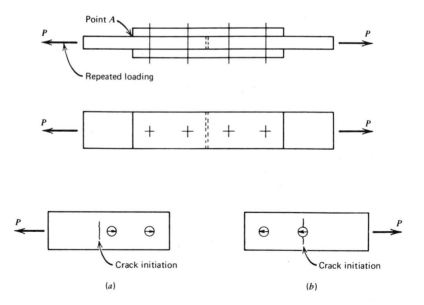

Fig. 5.37. Basic failure modes. (*a*) Gross section failure. (*b*) Net section failure.

completely transmitted by friction on the faying surfaces, implies a high concentration of shear stresses at point A. This results from the large differences in strain between the lap and main plates. The interface would be required to transmit a highly concentrated shear force at A if it were not relieved by microslip at that point. In many tests it was observed that, under these conditions, crack initiation and growth usually occurred in the gross section, in front of the first bolt hole, as indicated schematically in Fig. 5.37a. The cracks initiated on the faying surfaces of the connected plates. This phenomena is often referred to as fretting; it occurs at the interface between metallic surfaces that are in contact and that slip minute amounts relative to each other under the action of an oscillating force.[5.41] Even the small relative displacements between the lap and the main plates at point A (see Fig. 5.36) may be sufficient to initiate a fretting failure. The obvious effect of fretting is to damage the faying surfaces. Stress concentrations are also introduced, which in many cases lead to crack initiation and a further reduction in fatigue strength.

Tests have indicated that high contact pressures only exist in a small area around the bolt hole.[5.47,5.48] The normal stress due to the clamping force decreases rapidly from a maximum condition at the edge of the hole. The region where the normal stress acts depends on such geometrical factors as the plate thickness and bolt diameter. Usually, the circular pressure area falls within twice the diameter of the bolt. For this reason, the crack initiates at a section between the end of the lap plate and the bolt hole where the combination of microslip and normal pressure is more critical.

A typical fretting failure is shown in Fig. 5.38. Discontinuities of the mill scale,

Fig. 5.38. Typical fretting-type failure in gross section. (Courtesy of University of Illinois.)

the effective clamping zone of the bolt, and the frictional resistance all influence the point where fretting is initiated. Fretting is often apparent during fatigue testing. A powdery rust and mill scale dust usually works out from between the plates during testing.[5.20]

The other major type of fatigue failure that occurs in bolted or riveted shear-type splices is illustrated in Fig. 5.37b. The crack initiates at the edge of the hole and grows in the region of the net section. This condition occurs when most of the load is transmitted by shear and bearing, a situation that frequently develops in joints where the applied load exceeds the slip resistance of the faying surfaces. This results in higher net section stresses, and the edge of the hole becomes the point of crack initiation. Failure is brought about by fracture of the net section, as shown in Fig. 5.39

Both types of failure have been observed in tests. Often the two types of failure occur simultaneously in the same joint, as illustrated in Fig. 5.40. Final failure occurs partly through the net section and partly through the gross section.

Besides the bolt relaxation normally experienced after installation, some additional relaxation (5%) was observed during cyclic loading.[5.18] Tests have indicated that the total loss of bolt tension was rarely more than 10% of the initial bolt tension.[4.9,5.18,5.20]

5.3.2 Fatigue Strength of Bolted Butt Joints

The stress versus life relationship is best described by a logarithmic transformation of cycle life and maximum stress or stress range.[2.5-2.7] Therefore, data from fatigue tests are generally described using the relationship

$$\log N = A + B \log S_r \qquad (5.20)$$

where N represents the number of cycles, S_r the maximum stress or stress range, and A and B are constants. Plotted on a log-log scale, Eq. 5.20 results in a straight line. Work on both welded and mechanically fastened connections has suggested that knowledge of the type of detail present and the stress range to which it is subjected are sufficient to adequately describe the fatigue life (number of cycles) of steel structures.[5.56] The life is independent of the grade of steel used.

Since two basic types of crack growth were observed in bolted joints, one in the gross section and the other in the net section, the test results have been correlated with the stresses associated with both areas. If no major slip developed during the life of a specimen and high clamping forces are present, failure occurs in the gross section. Therefore, an examination of the test data using the stress range on the gross area seems reasonable. Net section stresses depend on geometrical factors such as the arrangement of the bolts in the joint. This causes large variation in stress and is partly responsible for the large scatter in test data when gross section failures are correlated on the basis of net section stresses. This is illustrated in Fig. 5.41 where test results from three different types of joints are compared.[5.5] The joint geometry is given in Fig. 5.42. Major slip did not occur

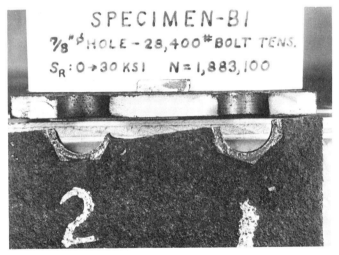

Fig. 5.39.   Failure at net section of bearing-type joint. (Courtesy of University of Illinois.)

because of the design conditions. Nearly all failures were through the gross section area. The test data indicate substantial variation in fatigue strength for the three different geometrical conditions. Figure 5.43 shows the same data plotted on the basis of the gross section stresses. These figures illustrate that the use of the gross area decreases the scatter in the test results significantly.

Major fatigue work on bolted and riveted connections was performed at the University of Illinois,[4.9,5.20,5.42] Northwestern University,[3.6,3.7,5.19] and in Ger-

Fig. 5.40. Crack initiation and growth at net section due to fretting. (Courtesy of U.S. Steel Corp.)

many.[5.18] Figure 5.44 shows some results of tests on bolted slip-resistant joints subjected to repeated loading. Since major slip did not occur in these joints, failure was caused by crack growth in the gross section. Therefore, the gross section area was used to determine the stress range, S_r, when evaluating the available test data. Most of the data were obtained from tests on steel specimens with a yield stress between 34 and 60 ksi. Data are available on joints fabricated from quenched and

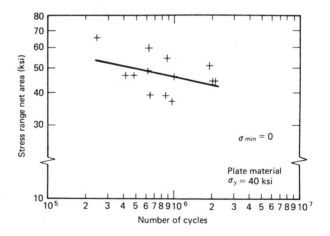

Fig. 5.41. Experimental S_r-N curve based on net area stress.

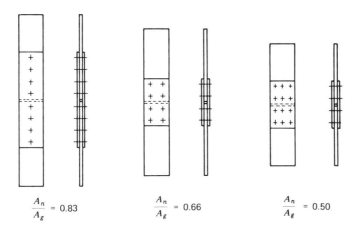

$$\frac{A_n}{A_g} = 0.83 \qquad \frac{A_n}{A_g} = 0.66 \qquad \frac{A_n}{A_g} = 0.50$$

Fig. 5.42.   Test specimen.

tempered alloy steel (A514) as well.[5.20] The yield stress (taken as the 0.2% offset) of the A514 steel was about 120 ksi. Although the data plotted in Fig. 5.44 show considerable scatter, they indicate that the yield stress of the material does not significantly influence the fatigue behavior of bolted joints.

Fatigue tests on slip-resistant joints in which the applied load on the specimen was reversed ($R < 0$) are shown in Fig. 5.45. The stress range includes the full compressive portion of the stress cycle. A comparison between the data plotted in Figs. 5.44 and 5.45 indicates that, for a given stress range, a slightly higher life was observed for the specimens subjected to stess reversal condition as compared with the zero-to-tension ($R = 0$) specimens. This seems reasonable in view of crack growth studies which indicate that, when residual tensile stresses are not present, the compression stress cycle is not as effective in extending the crack as

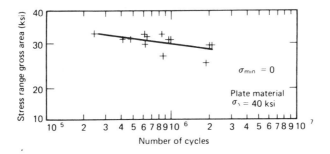

Fig. 5.43.   Experimental S_r-N curve based on gross area stress.

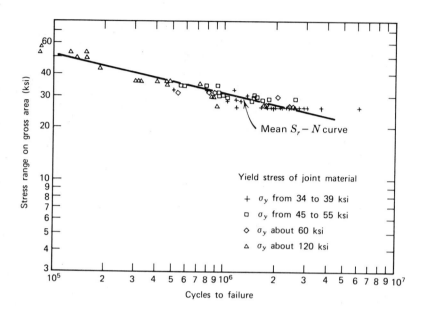

Fig. 5.44. Test results for slip-resistant joints ($R = 0$).

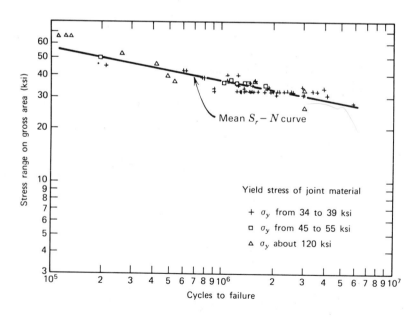

Fig. 5.45. Test results for slip-resistant joints ($R < 0$).

the tensile component.[5.43] Considering the full stress range effective results in a conservative estimate of the fatigue strength of bolted joints.

Joints with low slip resistance as a result of less clamping force or low slip coefficients are subjected to higher stresses on the net section if the slip resistance of the joint is exceeded by the applied load. When subjected to repeated loading conditions, crack initiation and growth occurs in the net section of such joints. Consequently, their performance under these loading conditions is related to the magnitude of stresses on the net section.

If the slip resistance of a joint is exceeded, the connection slips into bearing, and the applied load is transmitted partly by shear and bearing on the fastener as well as by friction on the faying surfaces. Tests have indicated that the fatigue life determined from a plate with a hole provides a lower bound estimate of the fatigue strength of bolted joints that have slipped into bearing.[5.18,5.19] The improved behavior of bolted joints as compared with the plate specimens with a hole is attributed primarily to the influence of the clamping force in the fasteners. A more favorable stress condition exists in the joint because part of the load is transmitted by friction on the faying surfaces, and because compressive stresses are introduced around the hole.

All available test results on bolted joints fabricated from steels with yield stress varying from 36 to 120 ksi are plotted in Fig. 5.46. Most of the results are from tension-type specimens or tension members, but a few are the results of tests on bolted cover plate ends in flexural members.[5.54] The stress range used to plot the test data was computed on the basis of the net or gross section area, depending on whether or not joint slip occurred. It is apparent from Fig. 5.46 that both bearing-type and slip-resistant joints subjected to reversal-type loading provide high fatigue strength.

The data plotted in Fig. 5.46 show a significant scatter even within the individual categories of joint types and loading conditions. This is mainly attributed to the fact that the data originated from various sources and reflected the variability in the hole fabrication, bolt clamping force, joint configuration, and other variables. Also, different tightening techniques were used to install the fasteners, and this may have resulted in significant variations in clamping forces of the fasteners. These variations as well as differences in joint geometry and hole preparation used in the various test series tend to increase the natural scatter of the data.

Except for a few tests on A514 steel joints, most of the test data obtained from axially loaded specimens were acquired at stresses that exceeded the yield point on the net section and often approached or exceeded the yield point on the gross section. However, the data from A514 steel joints are in good agreement with the other results in Fig. 5.46. The data from specimens that exceeded the yield point by large margins were not considered since they are not representative of the conditions that occur in actual structures.

Most of the data shown in Fig. 5.46 are concentrated in a stress range band between about 25 and 45 ksi. Additional information in the short life region and for a very large number of cycles is needed for a better understanding of the fatigue strength in these ranges.

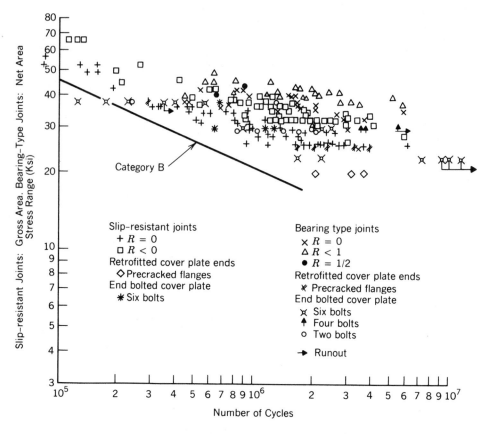

Fig. 5.46. Summary of test results of bolted joints.

Further examination of the data in Fig. 5.46 shows that both bearing-type and slip-resistant connections that are subject to load reversal ($R < 0$) have greater fatigue lives as compared with tension-only loading. Figure 5.47 is a plot of those cases from Fig. 5.46 in which load reversal occurs. However, the stress range used in Fig. 5.47 was taken as the algebraic difference between the maximum (tensile) stress and 60% of the minimum (compressive) stress. In this way, an attempt was made to recognize that in mechanically fastened joints not all of the compressive portion of the stress cycle is as damaging as the tension portion. This is because, unlike welded details, high tensile residual stresses are unlikely to be present in the bolted detail.

For design purposes, the data on bolted joints was compared with the 95% confidence limits used to define category B for fatigue specifications.[5.51] Although category B was derived from tests on plain welded beams, it was apparent that the

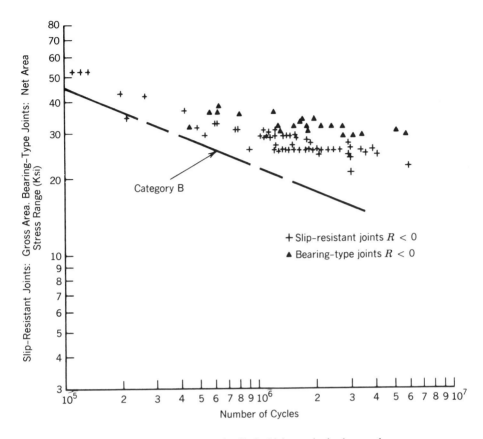

Fig. 5.47.   Fatigue strength of bolted joints under load reversal.

proposed design relationship provided a reasonable lower bound to the test data on bolted joints. The use of this lower bound for bolted joints results in conservative design relationships. It is apparent that slip-resistant joints designed on the basis of gross section and bearing-type joints designed on their net section provide about the same fatigue strength. For members under reversal of load, the use of a reduction factor, such as 60%, for the compression portion of the cycle can be used to reduce the amount of conservatism in selecting Category B.

The test data shown in Figs. 5.44 to 5.46 were developed from symmetric butt joints with a maximum of three fasteners in a line parallel to the direction of the applied load. Only a few test results of longer joints have been reported.[5.44] Tests of high-strength bolted joints with two, four, or six fasteners in a line indicated no significant influence of the number of bolts on the fatigue strength. These tests did show that the frictional resistance of the faying surfaces does affect the fatigue strength. An increase in slip resistance improved the fatigue behavior.

5.4 DESIGN RECOMMENDATION

5.4.1 Introduction

The mathematical model presented in Section 5.2 provides a reasonable prediction
of joint behavior under either working loads (allowable stress design) or factored
loads (load factor design). However, it is not suitable directly for design because
of its complexity. The results of analyses carried out using this "exact" solution
are used to form the basis for the design recommendations that follow. They are
presented in such a way that they can be used in either working stress design or
load factor design specifications.

Current design practice that is founded on an allowable stress format treats
mechanically fastened bearing-type joints on the basis of allowable stresses acting
on either the gross or net area of the member and on the average stresses in the
fasteners. Most design specifications do recognize, however, that the assumption
that each fastener carries an equal share of the load becomes less and less accurate
as joint length increases. The accommodation for this effect is generally applied in
a step-wise fashion (usually just one reduction in allowable shear stress with length),
although many European specifications provide a linearly varying reduction with
joint length. Specifications that use a load and resistance design format treat the
design of the fasteners in the same way; that is, the average fastener load is
generally applied, and the effect of joint length is recognized for longer joints.

In either case of allowable stress design or load factor design, once the member
forces are known from a structural analysis, the required number of fasteners can
be determined on the basis of the permissible shear stress for the fastener, including
consideration for the effect of joint length. Hence, the load transmitted by a bolted
joint with n fasteners and m possible shear planes per bolt through the bolt shank
can be expressed as

$$P = mn\tau_b A_b \qquad (5.21)$$

where τ_b represents the permissible shear stress on the fastener (allowable stress
design or load factor design, as appropriate), and A_b represents the nominal bolt
area. If the shear planes pass through the threaded parts of the bolt, Eq. 5.21 is
modified to

$$P = 0.70mn\tau_b A_b \qquad (5.22)$$

as discussed in Section 4.10.

The strength of a slip-resistant joint can be expressed in its most basic form,
Eq. 5.2, in which a slip coefficient is used. Alternatively, although the bolts are
not actually subjected to shearing forces, an equation such as Eq. 5.21 can be
used. In this format, an equivalent permissible shear stress is calculated, but it

must be remembered that the load is actually transferred by the frictional resistance on the faying surfaces.

Design criteria for connections can be based upon performance, strength, or both. In a slip-resistant joint, unsatisfactory behavior would result if a major slip occurred: a performance criterion. The function of the structure may be impaired due to misalignment or other unsatisfactory conditions that may result from the slip. However, most slip is minor and will not be detrimental to the performance of the joint. In these cases, strength is the factor that should govern the design; it is identified as the shear stress on the fastener, the bearing stress in the material adjacent to the fastener, or as the tensile stress on the net or gross cross-section of the member.

The ultimate capacity of both slip-resistant and bearing-type bolted joints is limited by failure of one or more components of the joint. Joint strength provides an upper bound for either joint type. Hence, in allowable stress design, the permissible strength of a slip-resistant joint can, at best, equal the capacity of an otherwise comparable bearing type connection. In other words, to design a slip-resistant joint, the slip resistance of the joint is determined on the basis of factors such as the surface condition, the bolt type, the tightening procedure, the number of bolts, and the number of slip planes. This slip resistance is then compared with the bolt shear capacity of the joint based upon the number of shear planes per bolt and their location (through the shank or through the threaded part of the bolt) and the number of bolts in the joint as well as the bolt quality. Of course, the smaller value of the shear strength and the slip resistance is governing.

In load factor design, the ultimate strength of the member or connection is checked against the effect of the factored loads. The factored load is determined by multiplying the working loads by a factor that is greater than 1.0. In addition, it is necessary for the member, joint, and structure as a whole to be ''serviceable'' at the working load level. This means that consideration must be given to control of deflections, deformation, and fatigue of the structure at its service or working load level.

To meet the requirements of load factor design, the ultimate strength of a bearing-type bolted joint is checked directly against the effect of the factored loads. Unless fatigue is a factor, the other requirements for serviceability are not operative since, by definition, any small slips that may occur are judged not to be detrimental.

On the other hand, a slip-resistant connection designed under load and resistance factor design must be checked under both service (working) load levels and factored load levels. The obvious requirement is that the connection not slip under working loads. In addition, however, it is still a requirement that the ultimate strength of the connection loads be checked under factored loads.

A connection that is subjected to fatigue loading must meet exactly the same requirements as those described for slip-resistant joints under either working stress design or load and resistance factor design, as appropriate. Of course, the governing permissible stress for the fatigue case is used to evaluate the resistance of the joint under the working loads.

5.4.2 Design Recommendations—Fasteners

i. Allowable Stress Design Bolted Joints. The balanced design concept has been used to develop design criteria for mechanically fastened joints in the past. This design philosophy results in wide variations in the factor of safety for the bolt because the ratio of the yield point to the tensile strength changes with various types of steel.[5.49] For example, the 1972 specifications[1.4] provided ratios of tensile strength to allowable tensile stress equal to 2.64, 2.48, and 2.00 for A36, A440, and A514 steels, respectively. Furthermore, the balanced design concept has no meaning when applied to long joints because the end fasteners may "unbutton" before the plate material can attain its full strength or before the interior bolts can be loaded up to their full strength. This "long joint" effect depends on the type of joint material as well as on the type of fastener.

All of these factors resulted in a variable factor of safety, as illustrated in Fig. 5.48. The factor of safety against failure of the joint is plotted as a function of joint length for several steels fastened with A325 bolts. An allowable shear stress of 22 ksi was used to proportion the fasteners. This was the allowable shear stress prescribed in the RCSC specification up to 1974. The allowable tensile stress on the net section of the joint was taken as 60% of the yield stress or 50% of the tensile strength of the plate material, whichever was smaller. It was apparent that a different approach is desirable; one that would provide both a rational method of determining the allowable stresses and a uniform, or at least a more consistent,

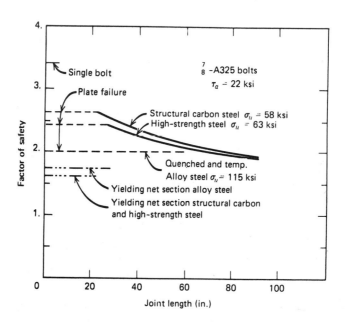

Fig. 5.48. Factor of safety versus joint length for A325 bolts.

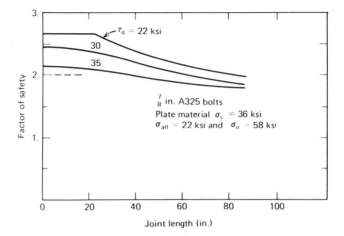

Fig. 5.49. Factor of safety for structural carbon steel joints fastened by A325 bolts.

factor of safety. It appeared that a more logical criterion to establish allowable stresses for the fasteners was to consider the fastener strength over the full range of joint behavior.

To determine the magnitude of the factor of safety deemed adequate for the fasteners, two aspects can be considered: (1) what the factor of safety has been in the past, and (2) what it ought to be. If past practice for riveted or bolted structural carbon steel joints is studied, the factor of safety against shear failure of the fastener is found to vary from approximately 3.3 for compact joints* to approximately 2.0 for joints with a length in excess of 50 in. This is illustrated in Fig. 5.48 for A325 bolts. The lower factor of safety for the longer joints was apparently adequate in the past. In fact, according to past practice, the largest and often most important joints have probably had the lowest factor of safety. Experience has shown that this factor of safety has provided a safe design condition. This indicated that a minimum factor of safety of 2.0 has been satisfactory; the same margin is also used for fasteners in tension. In addition, it was recognized that specified minimum mechanical properties of both the bolt and plate material were used to determine these lower bound conditions. Materials actually used as components of the joint usually provide strengths that exceed specified minimum properties. This results in an increased factor of safety. Finally, it can be noted that a minimum factor of safety equal to 2.0 for bolts in shear is not only in line with the factor of safety used for bolts in tension, but the same factor of safety against ultimate is also provided by quenched and tempered alloy steel tension members.[2.11]

---

*A compact joint is defined as a joint in which the average fastener shear stress at the ultimate load level is equal to, or almost equal to, the shear strength of a single fastener. The "unbuttoning" effect is negligible in these joints.

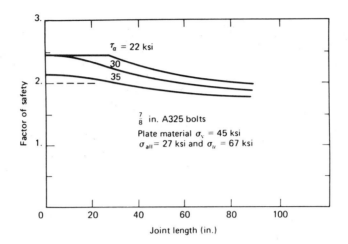

Fig. 5.50.   Factor of safety for high-strength steel joints fastened by A325 bolts.

In Fig. 5.49 the factor of safety is plotted as a function of the joint length for different allowable shear stresses in $\frac{7}{8}$-in. dia. A325 bolts, installed in structural carbon steel with a yield stress of 36 ksi and a tensile strength of 58 ksi. Joint length is defined as the length required to transfer the load from the main plate into the splice plates. Hence, for a symmetric butt splice, the joint length is equal to half the total length of the lap plate. For a single lap joint it is equal to the overall length of the joint. Figures 5.50, 5.51, and 5.52 show plots for other combinations of plate material and bolt grades. A minimum factor of safety of 2.0 is provided when the 30 ksi allowable shear stress is used for A325 bolts installed

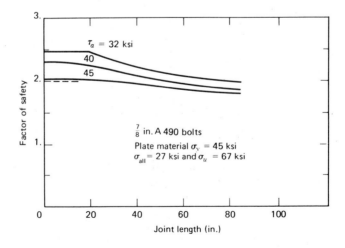

Fig. 5.51.   Factor of safety for high-strength steel joints fastened by A490 bolts.

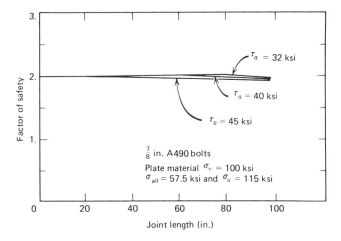

Fig. 5.52. Factor of safety for quenched and tempered alloy steel joints fastened by A490 bolts.

in structural carbon steel up to a joint length of 60 in. High-strength steel with a tensile strength of 66 ksi and fastened by A325 bolts provides a minimum factor of safety of 2.0 up to a joint length of about 50 in. Figures 5.51 and 5.52 show that a 40 ksi allowable shear stress for A490 bolts would provide the needed margin for joint lengths up to about 50 in. For joints with a length exceeding 50 in., the allowable shear stress in the bolts must be reduced to ensure a minimum factor of safety of 2.0. A 20% reduction in the allowable shear stress provides this margin for joint lengths between 50 and 90 in., as illustrated in Figs. 5.49 to 5.52.

Design Recommendations for Bolted Joints

ALLOWABLE STRESS DESIGN

Shear Stresses for High-strength Bolts

$$\tau_a = \beta \tau_{\text{basic}}$$

where τ_{basic} = 30 ksi — A325 bolts
τ_{basic} = 40 ksi — A490 bolts
β = 1.0 unless joint length exceeds 50 in., in which case β = 0.8.

Allowable Joint Loads

1. Shear planes pass through bolt shank

$$P = mn\tau_a A_b$$

2. Shear planes pass through bolt threads

$$P = 0.70mn\tau_a A_b$$

ii. Load Factor Design Bolted Joints. In load factor design, the connections and structural members are proportioned so that the product of maximum strength and a reduction factor ϕ is at least equal to the effect of the applied design loads multiplied by their respective load factors. The reduction factor ϕ is introduced to assure that the maximum strength of a structure is limited by the capacity of its members rather than by premature failure connections. The ϕ factor also accounts for the variability in strength of a connection. A uniform ϕ factor of 0.80 has been suggested for mechanical fasteners loaded in shear.[5.50]

The shear strength of a single fastener is about 60% of its tensile strength (see Section 4.2). A ϕ factor of 0.80 yields shear stresses comparable to those obtained by factoring the suggested working allowable shear values by 1.6. The same ϕ factor is applicable to A307 bolts and to A502 rivets. The ultimate shear capacity of a high-strength bolted connection is affected by the location of the shear planes. If a plane intersects the bolt threads, only the root area is effective in resisting the shear. This reduces the joint shear capacity by about 25% (see Section 4.10).

DESIGN RECOMMENDATIONS FOR BOLTED JOINTS

Load Factor Design—Shear Loading

$$\text{Design strength} = \phi F$$

where F − average shear strength $= 0.60\sigma_u$
ϕ − reduction factor $= 0.80$
If joint length exceeds 50 in. $\phi = 0.64$

Factored Joint Loads

1. Shear planes pass through bolt shank

$$P = mn\phi F A_b$$

2. Shear planes pass through bolt threads

$$P = 0.70mn\phi F A_b$$

iii. Slip-Resistant Joints. If it is assumed that equal clamping forces are present throughout a joint, then the slip resistance of a connection can be expressed as

$$P_s = mn T_i k_s \tag{5.23}$$

For a given joint geometry, the slip resistance is directly proportional to the product of the initial clamping force, T_i and the slip coefficient, k_s. Both quantities have considerable variance, and this must be considered when determining design criteria for slip-resistant joints. Since the frequency distributions for k_s and T_i are known for different surface conditions, bolt types, and tightening procedures (see Subsections 5.1.5 and 5.1.6), the joint frequency distribution for the product $k_s T_i$ can be determined[5.33] and suitable design expressions formulated. As an alternative to Eq. 5.23, an equivalent allowable bolt shear stress can be developed.

Considering Eq. 5.23, it will be desirable to reformulate this expression so that deterministic values can be used for T_i and k_s. Over and above this, it will be appropriate to provide design information for different levels of slip probability (the probability that the load predicted by Eq. 5.23 may be exceeded) in order that the designer might have the option of selecting a slip probability level suitable for his structure. Equation 5.23 can be written as

$$P_s = mn\alpha T_{i_{spec}} k_s \tag{5.24}$$

where

$$\alpha = T_i / T_{i_{spec}} \tag{5.25}$$

and $T_{i_{spec}}$ is the specified minimum bolt tension. In a further step, Eq. 5.24 will be expressed as

$$P_s = Dmn T_{i_{spec}} k_{s_{mean}} \tag{5.26}$$

where D is a multiplier that provides the relationship between $k_{s_{mean}}$ and k_s, incorporates α, and reflects the slip probability level selected.

The frequency distribution curve for the product of the two variables in Eq. 5.23, that is, T_i and k_s, is shown in Fig. 5.53a for A325 bolts fastening material in the clean mill scale condition and installed by the turn-of-nut method. Similar curves can be constructed for other fastener and faying surface conditions. A cumulative frequency curve constructed from this information is shown in Fig. 5.53b. If a very high value of $k_s T_i$, relative to the value actually present in the joint, were to be selected by the designer, then there would almost certainly be slip. On the other hand, if a very low value of $k_s T_i$ were selected as the design level, there would be very little likelihood of slip.

Two of the slip probability levels that might be chosen, 5% and 10%, are shown in Fig. 5.53b. The 5% slip probability (or 95% confidence level) corresponds to past practice for slip-resistant connections. If a lower slip probability is desired, the 1% level could be chosen; if a higher slip probability can be justified, 10% could be used.

Information like that given in Fig. 5.53b can be tabulated. Table 5.2 gives values of D for use in Eq. 5.26 for either A325 or A490 bolts installed by turn-of-nut and corresponding to various slip probability levels. The slip coefficients

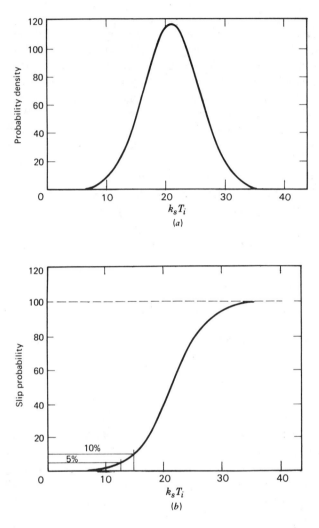

Fig. 5.53. Slip resistance. (*a*) Frequency distribution. (*b*) Cumulative frequency curve.

listed (mean values) are 0.20, 0.25, 0.33, 0.40, 0.50, and 0.60. The standard deviations used with these values in order to develop the table were 0.07 for mean values between 0.20 and 0.40 and 0.09 for the remainder. Table 5.3 gives similar information for A325 or A490 bolts installed using the calibrated wrench method.

A comparison of Tables 5.2 and 5.3 indicates that slip-resistant connections using bolt installed by the turn-of-nut method will have a slightly greater resistance than if the bolts were installed by calibrated wrench. For example, at the 5% slip-probability level, A325 bolts installed by turn-of-nut gain a premium of about 14% over A325 bolts installed by calibrated wrench. The difference reflects the higher preloads obtained in bolts installed by the turn-of-nut method. For A325 or A490

Table 5.2. Slip Factor D for use in Eq. 5.26: Turn-of-Nut Installation

	Slip Probability					
	A325 Turn-of-Nut			A490 Turn-of-Nut		
k_s (mean)	1%	5%	10%	1%	5%	10%
0.20	0.253	0.551	0.728	0.243	0.520	0.684
0.25	0.383	0.677	0.831	0.376	0.642	0.782
0.33	0.590	0.820	0.942	0.568	0.776	0.887
0.40	0.696	0.896	1.001	0.671	0.848	0.942
0.50	0.702	0.899	1.002	0.672	0.850	0.944
0.60	0.772	0.947	1.040	0.738	0.895	0.979

Note: Standard deviation of k_s (mean) taken as 0.07 for $k_s \leq 0.4$ and as 0.09 otherwise.

bolts installed by calibrated wrench, α is 1.13, whereas it is 1.35 for A325 bolts or 1.26 for A490 bolts installed by $\frac{1}{2}$ turn-of-nut, respectively.

The same information represented by Eq. 5.26 and Tables 5.2 and 5.3 can be expressed in terms of a permissible shear stress. (This is a convenience only; it must be remembered that the fastener in a slip-resistant connection is not actually acting in shear.) Equating the slip resistance (Eq. 5.3) to an equivalent shear force gives

$$mnT_i k_s = mn\tau_b A_b \qquad (5.27)$$

where τ_b is the equivalent shear stress and A_b is the nominal bolt area. Using α (Eq. 5.25) and expressing the specified bolt tension as

Table 5.3. Slip Factor D for use in Eq. 5.26: Calibrated Wrench Installation

	Slip Probability A325 or A490 Calibrated Wrench		
k_s (mean)	1%	5%	10%
0.20	0.235	0.478	0.622
0.25	0.372	0.594	0.714
0.33	0.547	0.718	0.810
0.40	0.639	0.784	0.862
0.50	0.643	0.787	0.864
0.60	0.702	0.829	0.897

Note: Standard deviation of k_s (mean) taken as 0.07 for $k_s \leq 0.4$ and as 0.09 otherwise.

$$T_{i_{\text{spec}}} = 0.7A_s \sigma_{u_{\text{spec}}} \tag{5.28}$$

where A_s is the stress area of the bolt, then Eq. 5.27 can be rewritten as

$$\tau_b = 0.7k_s \sigma_{u_{\text{spec}}} (A_s/A_b) \tag{5.29}$$

The ratio of the stress area to the nominal bolt area varies from only 0.736 for a $\frac{5}{8}$-in. diameter bolt up to 0.774 for a 1-in. diameter bolt. An average value of 0.76 will be used herein. The minimum specified tensile strength for A325 bolts in sizes $\frac{1}{2}$ through 1 in. diameter is 120 ksi. Substituting these values into Eq. 5.29 yields

$$\tau_b = 63.8k_s \alpha \tag{5.30}$$

Equation 5.30 relates the equivalent shear stress on the fastener to the known parameters α and k_s (as described in Section 5.1). An expression similar to Eq. 5.30 can be developed for A490 bolts; only the multiplier changes (to 78.7).

Of course, the frequency distribution and cumulative frequency distribution curves corresponding to Eq. 5.29 look just the same as those shown in Fig. 5.53a and b. Table 5.4 gives the equivalent permissible shear stresses for slip-resistant joints using A325 or A490 bolts installed by the turn-of-nut method, and Table 5.5 presents the same information for use when calibrated wrench installation is used. The slip coefficients selected ($k_{s_{\text{mean}}}$) and their standard deviations are the same as those used in Tables 5.2 and 5.3.

In evaluating conditions for A325 bolts, the specified minimum tensile strength was presumed to be 120 ksi. The specified tensile strength for A325 bolts in sizes over 1 in. diameter is in fact 105 ksi. Experience has shown that the actual strength of A325 bolts over 1 in. diameter usually ranges from 20 to 34% above the minimum specified tensile strength. Furthermore, the A_s/A_b ratio for these sizes is

Table 5.4. Equivalent Shear Stress for Use in Slip-Resistant Connections: Turn-of-Nut Installation

k_s (mean)	Slip Probability					
	A325 Turn-of-Nut			A490 Turn-of-Nut		
	1%	5%	10%	1%	5%	10%
0.20	3.23	7.03	9.29	3.82	8.18	10.77
0.25	6.11	10.80	13.25	7.40	12.63	15.39
0.33	12.42	17.27	19.84	14.74	20.16	23.03
0.40	17.75	22.85	25.53	21.12	26.70	29.66
0.50	22.39	28.67	31.98	26.44	33.44	37.14
0.60	29.56	36.24	39.79	34.82	42.28	46.23

Table 5.5. Equivalent Shear Stress for Use in Slip-Resistant Connections: Calibrated Wrench Installation

| | Slip Probability | | | | | |
| | A325 Calibrated Wrench | | | A490 Calibrated Wrench | | |
k_s (mean)	1%	5%	10%	1%	5%	10%
0.20	3.00	6.10	7.94	3.70	7.52	9.79
0.25	5.93	9.47	11.39	7.32	11.69	14.05
0.33	11.52	15.11	17.06	14.21	18.64	21.04
0.40	16.31	20.01	22.00	20.12	24.69	27.14
0.50	20.51	25.09	27.56	25.29	30.95	33.99
0.60	26.87	31.73	34.33	33.15	39.14	42.35

about 0.81 as compared with the value 0.76 for sizes less than 1 in. diameter. An increase in the A_s/A_b ratio increases the shear stress, as is apparent from Eq. 5.29. Hence, the values listed in Tables 5.2 through 5.4 are assumed applicable to all commonly used A325 bolt sizes.

A reduction factor must be applied to account for the effect of fabrication factors on the slip resistance of joints; for example, depending on the amount of oversize of the hole or the direction of the slotted holes with respect to the expected slip direction, a reduction in slip resistance may result. Chapter 9 deals specifically with oversize and slotted holes and discusses in greater detail the influence of these fabrication factors on the slip resistance of a joint.

Strength as well as performance must be considered in the design of slip-resistant joints. As mentioned in Subsection 5.4.1, the permissible load of a slip-resistant connection must not exceed its capacity based on considerations of strength. In other words, the permissible load for a joint evaluated on the basis of its strength capacity (as governed by shear of the bolts or bearing of the connected parts) forms the upper bound for the design of a slip-resistant connection. Slip-resistant connections governed by this upper bound are likely to be only those in which the slip coefficient is high or the probability of slip selected is high, or some combination of both of these. For example, a joint with a $k_{s_{mean}}$ value of 0.50 using A325 bolts installed by turn-of-nut will have a permissible equivalent shear stress of 32.0 ksi when designed against slip resistance. However, its capacity when checked as a bearing-type connection will be based on a permissible shear stress of only 30.0 ksi (Subsection 5.4.1). Thus, the latter governs even though this was a connection designed as slip resistant.

DESIGN RECOMMENDATIONS FOR SLIP-RESISTANT JOINTS

Slip-resistant joints may be proportioned in accordance with either Alternative A or Alternative B, as given below. The result will be the same in either case.

Alternative A

$$P_s = DmnT_{i_{spec}}k_{s_{mean}}$$

where D is obtained from Table 5.2 or 5.3

Alternative B

$$P_s = mn\tau_a A_b$$

where τ_a is obtained from Table 5.4 or 5.5 and A_b is the cross-sectional area corresponding to the nominal diameter of the bolt.

If slotted or oversize holes are used, the joint capacity calculated by either Alternative A or Alternative B must be reduced by multiplying by 0.70. See Chapter 9 for details on slotted and oversize holes.

In either allowable stress design or load factor design, the resistance described using either Alternative A or Alternative B is to be compared with the effect of the working loads (sometimes called specified loads in load factor design.) In allowable stress design, the slip-resistant joint must also be checked against its shear capacity (Subsection 5.4.2i) and its bearing capacity (Subsection 5.4.4i). In load factor design, the slip-resistant joint must likewise be checked against its shear capacity (Subsection 5.4.2ii) and its bearing capacity (Subsection 5.4.4ii) using factored loads.

5.4.3 Design Recommendations—Connected Material

It was noted in Section 5.2 that is was desirable that yielding through the gross cross-section of a member occur prior to failure at the net cross-section in order that the member behavior be ductile. That requirement is included in the design recommendations that follow. It includes a multiplier that reflects the fact that, while the actual yield and ultimate strengths can both be expected to be greater than their specified minimum values, the margin on yield is usually greater than that on ultimate.

i. Static Loading

a). Allowable Stress Design.   In allowable stress design, practice in the United States since 1978 has been to place a limit on the stress at the gross cross-section of the member, established at 60% of the yield strength of the material, and to require in addition that the stress on the net cross-section of the joint not be in excess of 50% of the tensile strength of the material. This provides a factor of safety of 1.67 against unrestricted plastic flow of the main member and a factor of safety of 2.0 against fracture. It will be recalled that the allowable shear stresses for bolts in bearing-type connections were established so that the factor of safety against fastener failure was at least 2.0. Thus, it can be expected that the tension member will reach its ultimate load prior to any (potential) failure of the bolts that make up its connection.

DESIGN RECOMMENDATIONS

Allowable stresses

Through gross cross-section of member,

$$\sigma_a = 0.60\ \sigma_y$$

or, through net cross-section at connection,

$$\sigma_a = 0.50\ \sigma_u$$

but,

$$\frac{A_n}{A_g} \geqslant \frac{\sigma_y}{0.9\ \sigma_u}.$$

Thus, the allowable load on the member is the lesser of

$$P_1 = 0.60\ \sigma_y A_g$$

or

$$P_2 = 0.50\ \sigma_u A_n.$$

b). Load Factor Design. The limit of strength of a tension member is its capacity as established by fracture at the net section. This capacity should be compared with the effect of the factored loads. A reduction (ϕ) will be applied to this nominal capacity ($A_n \sigma_u$) to reflect the possibility of undersize of member, accuracy of analysis, and actual material properties. For a safety index of 3.0, which is the value used for beams, columns, and beam-columns, a value of $\phi = 0.90$ is appropriate. It is worth noting that the safety index established for mechanically fastened connections[5.50] is 4.5, reflecting the desire that connections do not reach failure before the ultimate strength of the member has been attained.

In addition to strength, another limit state exists for tension members. This is unrestricted plastic flow of the main member, that is, yielding through the gross cross-section of the member. This could occur at loads only slightly greater than the working load level if only the strength limit were applicable. Thus, it is necessary that a second limit be applied, as noted below.

As was the case for tension members designed under the allowable stress method, the ductility of the member must also be ensured.

DESIGN RECOMMENDATIONS

Member capacity under factored loads shall be taken as the lesser of

$$P_f = \phi\ A_n \sigma_u$$

or

$$P_f = \phi \, A_g \sigma_y$$

where $\phi = 0.90$.
But

$$\frac{A_n}{A_g} \geq \frac{\sigma_y}{0.9 \, \sigma_u}.$$

ii. Repeated Loading. Results of fatigue tests on slip-resistant as well as other types of bolted joints were discussed in Section 5.3. It was shown that the type of failure was related to the manner in which the applied load was carried by the joint. If transmitted by frictional resistance on the contact surfaces alone, failure was through the gross section. When slip occurred and part of the load was transmitted by bearing and shear, failure generally occurred through the net section. The fatigue strength at the gross section of slip-resistant joints was about equal to the fatigue strength at the net section of joints that had slipped into bearing under nonreversible loading.

Design category B, which was derived from tests on plain welded beams,[5.51] provides a reasonable lower bound estimate for the stress range versus life relationship of bolted joints. The allowable stress ranges determined from this stress range versus life relationship for different loading conditions are summarized in Table 5.6. A stress range of 16 ksi was estimated for a life of 2 million cycles or more.

For the design of high-strength bolted joints under cyclic loading, the suggested stress range can be applied to: (1) the gross section area of slip-resistant joints with a slip probability of 5% or less, and (2) the net section area for other bolted joints. This provides design stresses for clean mill scale conditions that are in reasonable agreement with current practice. Joints subjected to reversal of stress should always be designed as slip-resistant joints in order to prevent excessive movement of the connected parts.

The stress range on the net section area governs the design of bolted joints that have a slip probability greater than 5%. These joints should not be used in situations

Table 5.6. Allowable Range of Stress for the Plate Material

Design Load Cycles	Stress Range for 95% Survival (ksi)
20,000–100,000	45.0
100,000–500,000	27.5
500,000–2,000,000	18.0
Over 2,000,000	16.0

where reversal of load occurs. However, slip in the direction of the maximum applied load is not critical unless the load is reversed.

Application of the stress ranges given in Table 5.6 provides a conservative design for both slip-resistant and bearing-type bolted joints. Better estimates of the stress range-life relationship may be developed when additional experimental data become available.

DESIGN RECOMMENDATIONS FOR JOINT MATERIAL UNDER REPEATED LOADING

Slip-Resistant Joints

Calculate stress range on gross section area if the slip probability is less than or equal to 5%.

Other Bolted Joints

Calculate stress range on the net area if the slip probability is greater than 5%. Stress reversal is not permitted. Allowable stress range for both types is given in Table 5.3

iii. Bearing Stresses. In Section 5.2.9 it was shown that the lower bound L/d ratio that prevents a single fastener from splitting out of the plate material can be expressed as:

$$\frac{L}{d} \geq 0.5 + 0.715 \frac{\sigma_b}{\sigma_u^P} \tag{5.31}$$

Butt joints with a single fastener were more critical than joints with multiple fasteners in a line. The clamping force in a high-strength bolt also has a favorable influence on the bearing strength of the connection. Hence, design recommendations based on test results of finger-tight single fastener specimens provide a conservative estimate of the required end distance.

The test results indicate that Eq. 5.31 provides an acceptable lower bound solution to the strength of the end zone for an L/d ratio up to 3.0 as illustrated in Fig. 5.54. When the L/d ratio exceeds 3.0, the failure mode changes gradually from a "shearing-type" failure to one in which large hole and material deformation occurs.

An alternative relationship can be used which directly relates the L/d ratio to the bearing stress-tensile strength ratio:

$$\frac{L}{d} \geq \frac{\sigma_b}{\sigma_u^P} \tag{5.32}$$

This relationship is also plotted in Fig. 5.54 and it is also in good agreement with the test data.

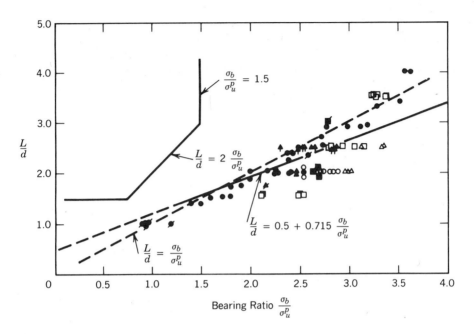

Fig. 5.54. Comparison of design recommendations for allowable stress design with test results.

a.   Allowable Stress Design.   If a minimum factor of safety with respect to ultimate load of 2.0 is selected, the required L/d ratio becomes

$$\frac{L}{d} \geq 0.5 + 1.43 \frac{\sigma_b}{\sigma_u^P} \tag{5.33}$$

As is shown in Fig. 5.54, Eq. 5.31 defines the L/d ratio up to a bearing stress-tensile strength ratio of about 3.0. The suggested factor of safety of 2.0 against bearing failure is comparable to the factors of safety against shear or tension failure of the fasteners and the tensile strength of the net section.

If the alternate formulation is used, the required L/d ratio becomes:

$$\frac{L}{d} \geq 2 \frac{\sigma_b}{\sigma_u^P} \tag{5.34}$$

To properly install a bolt or rivet, a minimum distance from the center of the fastener to any edge of the member must be maintained. A minimum L/d ratio of 1.5 is suggested since this conforms to current practice.

The design region shown in Fig. 5.54 is further bounded by a vertical line at a bearing stress-tensile strength ratio of 1.5. This prevents use of bearing stresses that may lead to excessive hole deformations and the upsetting of material in front of the fastener. Although the strength in such a situation is still adequate, large deformations may limit usefulness. Furthermore, a high σ_b/σ_u^P ratio corresponds to a large ratio of bolt diameter to the plate thickness. Thin plates that may deform out of their plane due to instability of the end section may limit the ultimate capacity of the end zone. These conditions may arise if the lap plates of a butt joint are critical in bearing. Due to "catenary action," the ends of the lap plates tend to bend outward. A high compressive force on the end panel may cause a dishing-type failure and decrease the ultimate bearing strength.

5.4.4 Design Recommendations for Bearing Stresses

i Allowable Stress Design

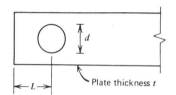

Bearing stress $\sigma_b = P/dt$
σ_u^P = tensile strength plate material

Following conditions are to be satisfied:

1. $L/d \geq 0.5 + 1.43\, \sigma_b/\sigma_u^P$; alternatively, $L/d \geq 2\sigma_b/\sigma_u^P$
2. $L/d \geq 1.5$
3. $\sigma_b/\sigma_u^P \geq 1.5$

ii *Load Factor Design.* A lower bound to the shear resistance of the end zone behind the fastener was expressed as (see Subsection 5.2.9):

$$F = (2t)\left(L - \frac{d}{2}\right)(0.7\,\sigma_u^P) \qquad (5.35)$$

A ϕ factor of 0.85 is believed adequate to account for the uncertainties in the strength of the end zone. Hence the shear strength of the end zone panel for load factor design becomes

$$\phi F = (0.85)\,(1.4)\left(L - \frac{d}{2}\right)t\sigma_u^P \qquad (5.36)$$

A minimum L/d ratio equal to 1.5 is desired for installation. In order to limit deformations of the hole, the bearing ratio σ_b/σ_u^P should not exceed 3.0 at the factored load level.

A ϕ factor of 0.85 provides bearing stresses on the fastener that are equal to those obtained by factoring the allowable bearing stress values given by Eq. 5.33.

DESIGN RECOMMENDATIONS FOR BEARING STRESSES

Load Factor Design

Shear strength end zone

$$F = (1.4) \left(L - \frac{d}{2} \right) t \sigma_u^P$$

Reduction factor $\phi = 0.85$
Following conditions are to be satisfied

1. Design load $\times$ load factor $\leq \phi F$; alternatively, $L/d \geq 1.7 \, \sigma_b/\sigma_u^P$
2. $L/d \geq 1.5$
3. $\sigma_b/\sigma_u^P \leq 3.0$

REFERENCES

5.1 W. H. Laub and J. R. Phillips, *The Effect of Fastener Material and Fastener Tension on the Allowable Bearing Stresses of Structural Joints*, Report 243.2, Fritz Engineering Laboratory, Lehigh University, Bethlehem, Pennsylvania, June 1954.

5.2 R. A. Hechtman, T. R. Flint, and P. L. Koepsell, *Fifth Progress Report on Slip of Structural Steel Double Lap Joints Assembled with High Tensile Steel Bolts*, Department of Civil Engineering, University of Washington, Seattle, February 1955.

5.3 R. A. Hechtman, D. R. Young, A. G. Chin, and E. R. Savikko, "Slip Joints Under Static Loads," *Transactions ASCE*, Vol. 120, 1955, pp. 1335–1352.

5.4 A. A. van Douwen, J. de Back, and L. P. Bouwman, *Connections with High Strength Bolts*, Report 6-59-9-VB-3, Stevin Laboratory, Department of Civil Engineering, Delft University of Technology, Delft, the Netherlands, 1959.

5.5 O. Steinhardt and K. Möhler, *Versuche zur Anwendung Vorgespannter Schrauben im Stahlbau, Teil II*, Bericht des Deutschen Ausschusses für Stahlbau, Stahlbau-Verlag Gmbh, Cologne, Germany, 1959.

5.6 G. H. Sterling and J. W. Fisher, "A440 Steel Joints Connected by A490 Bolts," *Journal of the Structural Division, ASCE*, Vol. 92, ST3, June 1966.

5.7 J. R. Divine, E. Chesson, Jr., and W. H. Munse, *Static and Dynamic Properties of Bolted Galvanized Structures*, Department of Civil Engineering, University of Illinois, Urbana, April 1966.

5.8 A Kuperus, *The Ratio Between the Slip Factor of Fe 52 and Fe 37, C.E.A.C.M. X-6-27*, Stevin Laboratory, Department of Civil Engineering, Delft University of Technology, Delft, the Netherlands, 1966.

5.9 G. C. Brookhart, I. H. Siddiqi, and D. D. Vasarhelyi, *The Effect of Galvanizing and Other Surface Treatment on High Tensile Bolts and Bolted Joints*, Department of Civil Engineering, University of Washington, Seattle, September 1966.

5.10 J. H. Lee and J. W. Fisher, *The Effect of Rectangular and Circular Fillers on the Behavior of Bolted Joints*, Report 318.6, Fritz Engineering Laboratory, Lehigh University, Bethlehem, Pennsylvania, June 1968.

5.11 J. H. Lee, C. O'Connor, and J. W. Fisher, "Effect of Surface Coatings and Exposure on Slip," *Journal of the Structural Division, ASCE*, Vol. 95, ST11, November 1969.

5.12 G. L. Kulak and J. W. Fisher, "A514 Steel Joints Fastened by A490 Bolts," *Journal of the Structural Division, ASCE*, Vol. 94, ST10, October 1968.

5.13 J. R. Divine, E. Chesson, Jr., and W. H. Munse, *Static and Dynamic Properties of Bolted Galvanized Structures*, Department of Civil Engineering, University of Illinois, Urbana, April 1966.

5.14 C. C. Chen and D. D. Vasarhelyi, *Bolted Joints with Main Plates of Different Thicknesses*, Department of Civil Engineering, University of Washington, Seattle, January 1965.

5.15 D. D. Vasarhelyi and K. C. Chiang, "Coefficient of Friction in Joints of Various Steels," *Journal of the Structural Division, ASCE*, Vol. 93, ST4, August 1967.

5.16 U. C. Vasishth, Z. A. Lu, and D. D. Vasarhelyi, "Effects of Fabrication Techniques," *Transactions ASCE*, Vol. 126, 1961, pp. 764–796.

5.17 M. Maseide and A. Selberg, *High Strength Bolts used in Structural Connections*, Division of Steel Structures, Technical University of Norway, Trondheim, Norway, January 1967.

5.18 K. Klöppel and T. Seeger, *Sicherheit und Bemessung Von H. V. Verbindungen Aus ST37 und ST52 Nach Versuchen unter Dauerbelastung und Ruhender Belastung*, Technische Hochschule, Darmstadt, Germany, 1965.

5.19 N. G. Hansen, "Fatigue Tests of Joints of High Strength Steels," *Journal of the Structural Division, ASCE*, Vol. 85, ST3, March 1959.

5.20 P. C. Birkemoe, D. F. Meinheit, and W. H. Munse, "Fatigue of A514 Steel in Bolted Connections," *Journal of the Structural Division, ASCE*, Vol. 95, ST10, October 1969.

5.21 J. W. Fisher and J. L. Rumpf, "Analysis of Bolted Butt Joints," *Journal of the Structural Division, ASCE*, Vol. 91, ST5, October 1965.

5.22 J. W. Fisher, "Behavior of Fasteners and Plates with Holes," *Journal of the Structural Division, ASCE*, Vol. 91, ST6, December 1965.

5.23 G. L. Kulak, "The Analysis of Constructional Alloy Steel Bolted Plate Splices," Ph.D. Dissertation, Lehigh University, Bethlehem, Pennsylvania, June 1967.

5.24 R. Kormanik and J. W. Fisher "Bearing Type Bolted Hybrid Joints," *Journal of the Structural Division, ASCE*, Vol. 93, ST5, October 1967.

5.25 J. W. Fisher and G. L. Kulak, "Tests of Bolted Butt Splices," *Journal of the Structural Division, ASCE*, Vol. 94, ST11, November 1968.

5.26 V. H. Cochrane, "Rules for Rivet Hole Deduction in Tension Members," *Engineering News-Record*, Vol. 80, November 16, 1922.

5.27 W. G. Brady and D. C. Drucker, "Investigation and Limit Analysis of Net Area in Tension," *Transactions ASCE*, Vol. 120, 1955, pp. 1133–1154.

5.28 W. H. Munse and E. Chesson, "Riveted and Bolted Joints: Net Section Design," *Journal of the Structural Division, ASCE*, Vol. 89, ST1, Part 1, February 1963.

5.29 E. Chesson and W. H. Munse, "Riveted and Bolted Joints: Truss-Type Tensile Connection," *Journal of the Structural Division, ASCE*, Vol. 89, ST1, Part 1, February 1963.

5.30 European Convention for Constructional Steelwork, *European Recommendations for*

Bolted Connections in Structural Steelwork, ECCS-T10-83-80, 4th Ed., Brussels, Nov. 1983.

5.31 W. H. Munse, *The Effect of Bearing Pressure on the Static Strength of Riveted Connections*, Bulletin 454, Engineering Experiment Station, University of Illinois, Urbana, July 1959.

5.32 J. Jones, "Bearing-Ratio Effect on Strength of Riveted Joints," *Transactions ASCE*, Vol. 123, 1958, pp. 964–972.

5.33 L. A. Aroian, "The Probability Function of the Product of Two Normally Distributed Variables," *Annals of Mathematical Statistics*, Vol. 18, p. 265, 1947.

5.34 K. C. Chiang and D. D. Vasarhelyi, *The Coefficients of Friction in Bolted Joints Made with Various Steels and with Multiple Contact Surfaces*, Department of Civil Engineering, University of Washington, Seattle, February 1964.

5.35 R. A. Bendigo, R. M. Hansen, and J. L. Rumpf, *A Pilot Investigation of the Feasibility of Obtaining High Bolt Tensions Using Calibrated Impact Wrenches*, Report 200.59. 166A, Fritz Engineering Laboratory, Lehigh University, Bethlehem, Pennsylvania, November 1959.

5.36 J. de Back and L. P. Bouwman, *The Friction Factor Under Influence of Different Tightening Methods of the Bolts and of Different Conditions of the Contact Surfaces*, Stevin Laboratory, Report 6-59-9-VB-3, Delft University of Technology, Delft, the Netherlands, August 1959.

5.37 S. Hojarczyk, J. Kasinski, and T. Nawrot, "Load Slip Characteristics of High Strength Bolted Structural Joints Protected from Corrosion by Various Sprayed Coatings," *Proceedings, Jubilee Symposium on High Strength Bolts*, the Institution of Structural Engineers, London, 1959.

5.38 G. L. Kulak, "The Behavior of A514 Steel Tension Members," *Engineering Journal AISC*, Vol. 8, No. 1, January 1971.

5.39 J. de Back and A. de Jong, *Measurements on Connections with High Strength Bolts, Particularly in View of the Permissable Arithmetical Bearing Stress*, Report 6-68-3, Stevin Laboratory, Delft University of Technology, Delft, the Netherlands, 1968.

5.40 M. Hirano, "Bearing Stresses in Bolted Joints," *Society of Steel Construction of Japan*, Vol. 6, No. 58, Tokyo, 1970.

5.41 K. L. Johnson and J. J. O'Connor, "Mechanics of Fretting," *Proceedings of the Institution of Mechanical Engineers*, Vol. 178, Part 3J, London 1963–1964.

5.42 P. C. Birkemoe and R. S. Srinivasan, "Fatigue of Bolted High Strength Structural Steel," *Journal of the Structural Division, ASCE*, Vol. 97, ST3, March 1971.

5.43 T. R. Gurney, "The Effect of Mean Stress and Material Yield Stress on Fatigue Crack Propagation in Steel," *Metal Construction and British Welding Journal*, Vol. 1, No. 2, February 1969.

5.44 J. Tajima and K. Tomonaga, *Fatigue Tests on High-Strength Bolted Joints*, Structural Design Office, Japanese National Railways, Tokyo, 1963.

5.45 E. Chesson, Jr., "Bolted Bridge Behavior During Erection and Service," *Journal of the Structural Division, ASCE*, Vol. 91, ST3, June 1965.

5.46 A. Nadai, *Theory of Flow and Fracture of Solids*, Vol. 1, 2nd ed., McGraw-Hill, New York, 1950, p. 229.

5.47 I. Fernlund, *A Method to Calculate the Pressure Between Bolted or Riveted Plates*,

Report 17, Inst. Machine Elements, Chalmers University of Technology, Gothenburg, Sweden, 1961.

5.48 J. W. Carter, K. H. Lenzen, and L. T. Wyly, "Fatigue in Riveted and Bolted Single-Lap Joints," *Transactions ASCE*, Vol. 120, 1955.

5.49 J. W. Fisher and L. S. Beedle, "Criteria for Designing Bearing-Type Bolted Joints," *Journal of the Structural Division, ASCE*, Vol. 91, ST5, October 1965.

5.50 J. W. Fisher, T. V. Galambos, G. L. Kulak, and M. K. Ravindra, "Load and Resistance Factor Design Criteria for Connectors," *Journal of the Structural Division ASCE*, Vol. 104, ST9, September, 1978.

5.51 J. W. Fisher, P. Albrecht, B. T. Yen, D. J. Klingerman, and B. M. McNamee, *Fatigue Strength of Welded Beams*, NCHRP Report 147, Highway Research Board National Academy of Sciences, 1974.

5.52 W. H. Munse, Addendum to Preliminary Report on Short-Grip High-Strength Bolts, Department of Civil Engineering, University of Illinois, Urbana, February 1974.

5.53 K. H. Frank and J. A. Yura, An Experimental Study of Bolted Shear Connections, Report No. FHWA/RD-81/148, Federal Highway Administration, U.S. Department of Transportation, Washington, D.C., December 1981.

5.54 A. H. Sahli, P. Albrecht, and D. W. Vannoy, "Fatigue Strength of Retrofitted Cover Plates," *Journal of the Structural Division, ASCE*, Vol. 110, No. 6, June 1984.

5.55 J. A. Yura, K. H. Frank, and L. Cayes, "Bolted Friction Connections with Weathering Steel," *Journal of the Structural Division, ASCE*, Vol. 107, ST11, November 1981.

5.56 J. W. Fisher, *Bridge Fatigue Guide*, American Institute of Steel Construction, Chicago, 1977.

Chapter Six

Truss-Type Connections

6.1 INTRODUCTION

Chapter 5 summarized the strength, behavior, and design of flat plate joints. The common features of such joints are that (1) all shear planes in the joint are parallel to one another and (2) all material in the joint is adjacent to a shear plane. In practice, most structural members do not consist solely of plates; they may be single rolled shapes, combinations of rolled shapes, or combinations of rolled shapes and plates. If the joints are to be fabricated by means of welding, it may be possible to connect one member directly to another. However, when bolted connections are used, it is usually necessary to transfer the load by means of gusset plates. Figure 6.1a shows how a light built-up member consisting of two angles uses a gusset plate in one plane to transfer the load out of the member. In Figure 6.1b two gusset plates are used at the end of a single rolled shape to transfer the load. (These cases are more fully depicted in Fig. 2.4c.) Note that in each situation the amount of material directly connected is less than 100% of the amount of the main member cross-sectional area.

In comparing the types of connections shown in Fig. 6.1 with the flat plate joints discussed in Chapter 5, it is evident that two distinct differences have been introduced; the amount of connected material may be less than 100% of the area of the main member, and a new component is present, namely, the gusset plate. In this chapter, the strength of the member as affected by this type of connection will be discussed. The design and behavior of the gusset plate is examined in Chapter 15.

6.2 BEHAVIOR OF TRUSS-TYPE CONNECTIONS

6.2.1 Static Loading

Members in truss systems are subjected to either compression or tension forces. Unless buckling governs, tension is more critical because the strength of both the member and its connection will be governed by the net cross-sectional area. In the discussion that follows, only members loaded in tension will be considered.

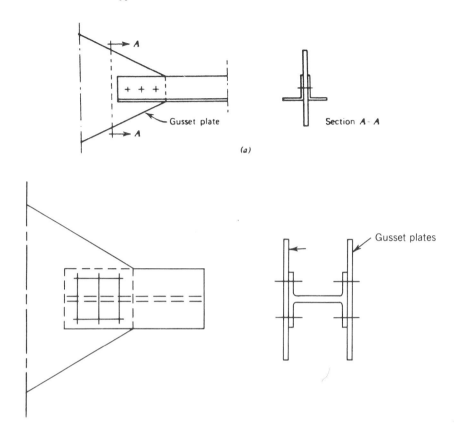

Section A - A

(a)

Gusset plate

Gusset plates

Fig. 6.1. (a) Single and (b) double gusset plate connection.

In butt splices made for plate members, the centers of gravity of the forces in the splice plates are close enough to the center of gravity of the force in the main member that they can be considered coincident. In members of the types shown in Fig. 6.1, this assumption is no longer valid, and the effect of the distances between the centers of gravity of the various components must be taken into account. In the case of the member built up from angles (Fig. 6.1a), the effect is that resulting from the distance between the centroid of each of the angles and the centroid of the gusset. In the single component member in which the force is taken out by two gusset plates (Fig. 6.1b), it is the effect of the displacement of the centroids of the gusset plates with respect to that of the main member.

Tests have indicated that the net section efficiency of members in which less than the total cross-sectional area of the member is connected shows a significant variance.[5.28,5.29,6.3] In addition to the non-alignment of centroids, other variables affect this efficiency. These include the ductility of the material being joined, the method of making the bolt holes (punched, subpunched and reamed, or drilled), and the ratio of hole gage to hole diameter. However, it is the position of the shear planes relative to the various parts of the cross-section of the member that has the greatest influence.[5.28,5.29]

All member components are assumed to be uniformly stressed at some distance from the connection region. Measurements have shown this to be a reasonable assumption.[6.4] However, for members like those shown in Fig. 6.1, a nonuniform stress distribution is created in the connection region because not all member components are connected to the gusset plates. For instance, whatever load is in the outstanding leg of the angles shown in Fig. 6.1a must be transferred through the fasteners placed in the other leg of the angle. Similarly, the load in the web of the member shown in Fig. 6.1b must be transferred to the gusset through the fasteners in the flanges. This generally results in higher stresses in the components that are attached directly to the gusset plates. Depending on joint geometry and material characteristics, this may result in a decrease in efficiency of the net section in the connection region because these components tend to reach their ultimate strength before the complete net section capacity has been developed. Similar results were observed in tests of angles welded to a gusset plate.[6.5] This loss of efficiency as a result of the distribution of cross section material relative to the gusset plate is referred to as "shear lag."

Munse and Chesson have examined the tensile behavior of various cross sections. They observed that the loss in efficiency at the net section due to shear lag was related to the ratio of the length L of the connection and the eccentricity $\bar{x}$ from the face of the gusset plate to the center of gravity of the connected component (see Fig. 6.2a).[5.28,5.29] The parameter $\bar{x}/L$ accounts for the effectiveness of the cross-section material with respect to the shear plane between the member and the gusset plate. The significance of this factor is discussed hereafter.

The unequal distribution of fastener loads in a butt joint was discussed in Chapter 5. A similar load distribution occurs among the fasteners in joints of the type shown in Fig. 6.1. Hence, relatively high loads are transferred by the end fasteners. As a result, fastener failures similar to the ones observed in long symmetric butt joints have been observed in members connected by gusset plates as well.[5.28,5.29,6.3]

The length L of the connection not only affects the load distribution among the fasteners but also influences the shear lag in a connection. Munse and Chesson concluded that a decrease in joint length increases the shear lag effect. This conclusion was based on test results from connections of the type as shown in Fig. 6.2b, which were tested to failure with either 5 or 10 A325 bolts in line in the connection region.[5.29] In either case, failure of the members occurred in the net section at the first line of fasteners, as illustrated in Fig. 6.3. The member with five bolts in a line had less strength (about 18%) at the net section compared with the longer joint with 10 bolts in a line. The fasteners were not the critical components for either test joint. Since the geometry of both joints was the same except for the joint length, it was concluded that the efficiency of the net section increases with a decrease in the ratio of $\bar{x}/L$.[5.28,5.29] Hence, an increase in joint length generally increases the effectiveness of the net section but decreases the effectiveness of the fasteners.

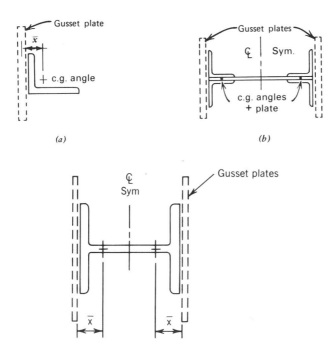

Fig. 6.2. Schematic of eccentricity in joints.

To approximate the efficiency of the net section by taking into account joint length and joint geometry, Munse and Chesson suggested that the actual net area be reduced to an effective net section area by applying a reduction factor V to account for the shear lag.[5.28,5.29] The reduction factor V was defined by the following empirical relationship

$$V = 1 - \left(\frac{\bar{x}}{L}\right) \tag{6.1}$$

where L is the joint length and $\bar{x}$ is the eccentricity between the shear plane and the centroidal axis of the connected component (see Fig. 6.2). Hence, the effective net section area of a built-up member is equal to

$$\text{effective area} = A_n\left(1 - \frac{\bar{x}}{L}\right) \tag{6.2}$$

where A_n is the net area of the connected member, calculated in accordance with Eq. 5.11 (the $s^2/4g$ rule). The definition of $\bar{x}$ is generally straightforward and is illustrated in Fig. 6.2. It should be noted that for rolled shapes (Fig. 6.2c) or built-

Fig. 6.3. Angle failure in built-up section. (Courtesy of University of Illinois.)

up shapes (Fig. 6.2*b*), the distance $\bar{x}$ is to be referred to the center of gravity of the material lying to either side of the centerline of symmetry of the cross section. In the case illustrated in Fig. 6.2b, this center of gravity would be that of two angles and one-half of the web plate. For the member shown in Fig. 6.2c, the "connected component" is equivalent to a T-section.

Although shear lag is the major factor that reduces the efficiency of the net section, it has been noted that other factors such as ductility of the material, the ratio of the fastener gauge g to the fastener diameter d, and fabrication procedures also influence the efficiency of the net section. In addition, Fig. 5.28 showed that the A_n/A_g ratio influences the tensile strength of the material of planar tension specimens. Generally, an increase in tensile strength accompanied a decrease in A_n/A_g.

In the case of members for which less than the entire cross-section is connected to gusset plates, only the connected portions are subject to a variation in strength with changes in the A_n/A_g ratio. Therefore, the influence of the A_n/A_g ratio on the net section of the member is less pronounced than in butt-type connections joining plate elements.

Ductility of the member material affects the net section strength as well as the load distribution among the fasteners. An increase in ductility tends to increase the net section strength and provides a more uniform load transfer among the fasteners.

It was pointed out in Section 2.7 that punched holes should be reamed to remove the work-hardened material that exhibits low ductility and may contain small cracks as a result of the fabrication process. For these reasons, joints with punched holes often show a decreased efficiency when compared with similar sections with punched and reamed holes or drilled holes. This condition can be more critical if substantial shear lag exists as well.[5.28]

Munse and Chesson developed empirical relationships to account for these factors mentioned above. They first compared the observed efficiency of test data with the efficiency of a member computed on the basis of the net section without accounting for the influence of factors such as shear lag, and such. As expected, a significant scatter of the data resulted, as shown in Fig. 6.4a. The scatter of data was significantly reduced when the observed test efficiency was compared with a computed efficiency that accounted for such factors as shear lag, ductility of the

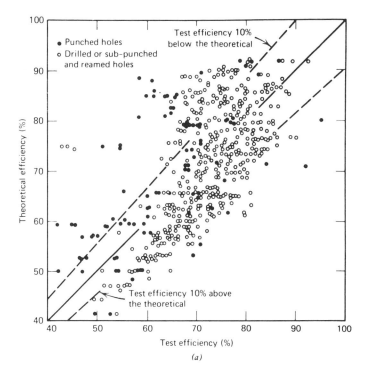

(a)

Fig. 6.4. Correlation of theoretical and test efficiencies. (a) Based on net area. (b) Based on effective net area.

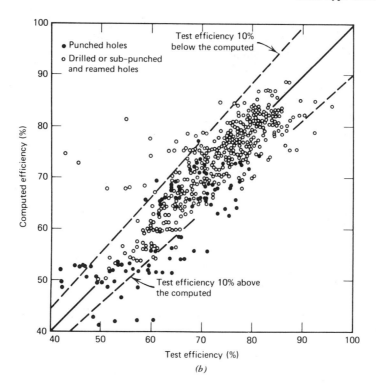

Fig. 6.4. (*Continued*)

material, fastener spacing, and fabrication procedure. This is illustrated in Fig. 6.4*b*. They concluded that, for most of the connections examined, shear lag was the major factor causing the difference between experimental and predicted efficiencies. The application of Eq. 6.2 presents some difficulties in practice. Given the member load, the designer must first select a cross-section on a trial basis and then design the connection in order to use Eq. 6.2. After establishing the effective net area according to Eq. 6.2, he must again examine the trial section to see if it is adequate. Several iterations may be required. After an extensive examination of the test data (more than 1000 tests are represented in Fig. 6.4) and many more hypothetical cases, it was concluded by a Task Committee of AISC that a simpler specification of effective net area was possible. These rules, contained in the current AISC specification,[2.11] are as follows:

$$A_e = C_t A_n \qquad (6.3)$$

where A_n is the net area calculated according to the $s^2/4g$ rule and C_t is a reduction coefficient given as follows:

1. $C_t = 0.90$ for W, M, or S shapes with flange widths not less than 2/3 the

depth, and structural tees cut from these shapes. Connection must be made to the flanges and there must be no fewer than three fasteners per line in the direction of the force in the member.

2. $C_t = 0.85$ for W, M, or S shapes that do not meet the requirements of 1, structural tees cut from these shapes, and all other shapes, including built-up cross-sections. Connection must be made to the flanges, and there must be no fewer than three fasteners per line in the direction of the force in the member.

3. $C_t = 0.75$ for all members whose connections have only two fasteners per line in the direction of the force in the member.

In the case described by 1, the flange area predominates; $\bar{x}$ is therefore relatively small, and the efficiency, $(1 - \bar{x}/L)$, is relatively high. Case 2 covers the range 0.67–0.90 of $(1 - \bar{x}/L)$ and is the mean value. In both cases 1 and 2 it is required that there be at least three fasteners per line in the direction of the force in the member. If this requirement is not met, the shear lag is more severe, and a lower value of C_t is provided for all cross-sections for this case 3.

6.2.2 Repeated Loading

The fatigue strength of built-up structural shapes, especially in the connection region, has been the concern of many engineers as experience has shown this to be a critical factor for repeatedly loaded structures. Several failures of riveted members in truss bridges constructed of built-up sections were attributed to fatigue.[6.4] A detailed analysis of all the factors involved is not possible, but some guidance can be obtained from an examination of these failures.

A survey of the fatigue failures observed in riveted bridges showed that the fatigue cracks in members often initiated from the side of a rivet hole at the edge of the gusset plate or splice plates (see Fig. 6.5). When cracks occurred in the gusset plate, they started at the sides of the rivet holes at the end of the members, as indicated in Fig. 6.5. Severe stress concentrations provided by geometry and shear lag in combination with the initial flaw conditions at those points made those

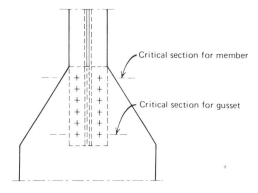

Fig. 6.5. Critical sections for a joint in a built-up section subjected to fatigue loading conditions.

locations susceptible to crack growth. The initial flaw condition for these joints is basically not different from conditions encountered in other bolted or riveted splices. Small microcracks at the sides of the hole are present as a result of the fabrication process. The stress concentration in connections of built-up or rolled shape members is likely to be more severe than encountered in symmetric butt splices because of the resulting eccentricities and shear lag. This is more severe in riveted joints, because the clamping force is not as great as in bolted joints, and more localized bearing occurs. Stress concentrations at the end rivet holes are further aggravated by the unequal load distribution among the fasteners. Sometimes these conditions may not significantly influence the static strength of the connection, but they may adversely affect the fatigue strength.

The fatigue strength is improved when rivets are replaced by high-strength bolts. This procedure has been used to overcome fatigue-related problems in existing riveted bridge joints. The high clamping force in the bolt results in a much better stress condition at the critical sections at the fastener holes. If sufficient slip resistance is provided, bearing stresses are eliminated and crack initiation and growth is not as critical at fastener holes.

Because of symmetry and the existence of a web plate, the connections shown in Fig. 6.1 do not develop severe secondary stresses from out-of-plane bending. When eccentrically loaded members are used and these secondary deformations are not prevented by proper lacing or diaphragms, the member tends to align, and this results in additional bending stresses. Although the static strength is not greatly affected,[6.2,6.5] severe reductions in fatigue strengths have been observed.[6.2] Net section as well as gross section fatigue failures developed prematurely in eccentrically loaded members and depended on the loading at the joint geometry. Reductions in life up to 80% were observed when compared with the data obtained from tests on similar symmetric butt splices.[6.2] This reduction is due to severe stress conditions caused by the secondary stresses resulting from out-of-plane deformations. These tests indicated clearly the need for proper restraints of the connection if the possibility of fatigue failure is to be minimized. When restraints to out-of-plane bending are provided, the fatigue strength of bolted connections in built-up truss members is comparable to the fatigue strength of similar butt joints.

6.3 DESIGN RECOMMENDATIONS

The design recommendations given in Section 5.4 for bolts in slip-resistant and bearing-type joints are also applicable in those cases where the shear planes of the various components are not coincident with the centers of gravity of the connected parts. Although the load distribution among the fasteners in joints for built-up sections is not identical to plate butt splices, the difference is considered negligible for practical purposes.

The static strength of the net section of a tension member was shown to be affected by several factors.[5.28,5.29] However, the dominant factor has been shown to be the influence of shear lag, and its significance should be considered in the design process. The empirical formula proposed by Chesson and Munse[5.28,5.29] provides a reasonable way of calculating the effective net area. This formula is

$$A_e = A_n \left(1 - \frac{\bar{x}}{L}\right) \qquad (6.4)$$

After calculating the effective net area, the rules provided in subsection 5.4.3 for member design can be used, replacing the net area in the design equations by the effective net area. If desired, the simplified net area rules given in the AISC specification[2.11] and described in Section 6.2 can be used in place of Eq. 6.4.

Present AASHTO specifications incorporate shear lag effects in tension members consisting of single angles or T-sections by assuming the effective net section area to be equal to the net area of the connected leg or flange plus one-half of the area of the outstanding leg.[2.2] Additional requirements regarding the effective net section are provided for some other joint geometries. These requirements have greater applicability when members are subjected to cyclic loading.

When fatigue is to be considered in the design of a joint or net area for a built-up section, sufficient restraints should be provided to prevent secondary stresses from developing. Slip-resistant joints are preferred for high fatigue strength. The deisgn recommendations given in Chapter 5.4 for butt-type joints are applicable to these types of joints when secondary stresses are minimized. The governing net section stress should be evaluated on the basis of an effective net section in order to account for the stress raising effects due to shear lag and other factors.

REFERENCES

6.1 AREA Committee on Iron and Steel Structures, "Stress Distribution in Bridge Frames-Floorbeam Hangers," *Proceedings, American Railway Engineering Association*, Vol. 51, 1950, pp. 470-503.

6.2 K. Klöppel and T. Seeger, "Dauerversuche Mit Einschnittigen HV-Verbindugen Aus ST37," *Der Stahlbau*, Vol. 33, No. 8, August, and No. 11, October 1964.

6.3 E. Chesson, Jr., and W. H. Munse, "Behavior of Riveted Truss Type Connections," *Transactions, ASCE*, Vol. 123, 1958, pp. 1087-1128.

6.4 L. T. Wyly, M. B. Scott, L. B. McCammon, and C. W. Lindner, *A Study of the Behavior of Floorbeam Hangers*, American Railway Engineering Association Bulletin 482, September, October 1949.

6.5 G. J. Gibson and B. T. Wake, "An Investigation of Welded Connections for Angle Tension Members," *Journal of the American Welding Society*, Vol. 7, No. 1, January 1942.

Chapter Seven

Shingle Joints

7.1 INTRODUCTION

In contrast to butt-type splices, the main components of the members of shingle joints are spliced at various locations along the joint. By terminating the main plates at different locations, the continuation plate can also serve as a cover plate over several regions of the joint (see Fig. 7.1). This type of connection provides a more gradual transfer of load in the plates throughout the joint. The connection is often used where the main member consists of several plies of material. Typical examples are the built-up box sections of chord members of truss bridges.

Shingle joints result in less joint thickness than butt joints, since butt joint requires all the force to be transferred into the lap plates. In a shingle joint the load is carried by the lap plates as well as by the continuous main plates at each plate discontinuity. Shingle joints can also facilitate the connection of various bridge components in a truss bridge. For example, plate A in Fig. 7.1 may also serve as a gusset for other members framing into the chord.

Shingle joints are most often used where reversal of stress is unlikely to occur because of the large dead load. Hence, most shingle joints are not slip-critical, and joint strength, rather than slip, is the governing criteria. Because special situations may require a design to be slip resistant, design recommendations for both types of load transfer are given.

7.2 BEHAVIOR OF SHINGLE JOINTS

Figure 7.2 shows a typical load versus deformation curve for a shingle joint.[7.1] This particular joint consisted of three regions with six $\frac{7}{8}$-in. dia. A325 bolts in each region. The plates had a clean mill scale surface condition and the yield strength of the plate material was about 50 ksi. The load versus deformation curve shown in Fig. 7.2 indicates that in the early load stages the load is completely carried by the frictional forces acting on the faying surfaces. Tests have demonstrated that shingle joints often exhibit two distinct load levels at which major slip occurs. At the first slip load, movement develops mainly along the shear plane

158

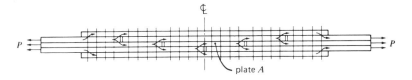

Fig. 7.1. Force flow in typical triple plate shingle joint.

adjacent to the main plate terminations. This slip plane is depicted as plane A in Fig. 7.2. At first, little slip or no movement is observed along the second slip plane, indicated as plane B in Fig. 7.2. Upon increasing the load, a second major slip occurs, with slip developing along the second slip plane (plane B in Fig. 7.2). At the same time some additional slip develops along the first slip plane (plane A).

It has been observed in tests on shingle joints that the total amount of slip tends to be less than the hole clearance.[7.1,7.5] This is especially true for large and complex bolted joints, mainly because of unavoidable misalignment tolerances during the fabrication process.

After major slip, the behavior of shingle joints is in many respects similar to the behavior of symmetric butt joints. Because the fasteners are bearing against the plate material, fastener deformations are developed in proportion to the load transmitted by each fastener. At high load levels the load versus deformation relationship

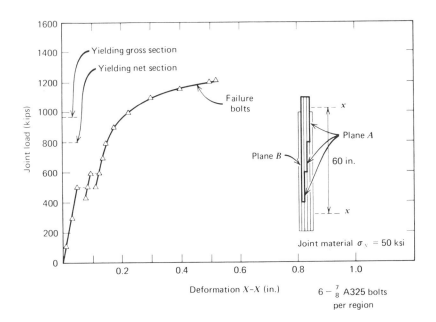

Fig. 7.2. Load versus deformation behavior of shingle joint.

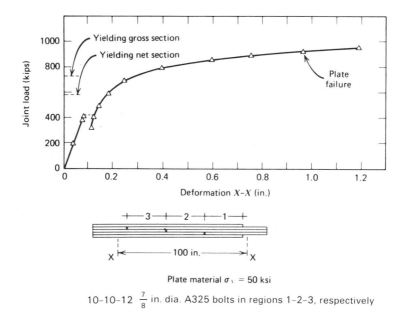

Fig. 7.3. Load versus deformation behavior of shingle joint.

of the joint becomes nonlinear because of plastic deformations in the fasteners and the plates. Depending on the joint geometry and the mechanical properties of the constituent parts, failure occurs either by shearing of the fasteners or by fracture of the plates. Both types of failures have been experienced in tests.[7.1,7.5] Characteristic load versus deformation curves are shown in Figs. 7.2 and 7.3.

Although both shingle and symmetric butt joints yield similar load versus deformation relationships, the deformation pattern of the individual fasteners is usually quite different. This is illustrated in Fig. 7.4 where a sawed section of a three-region joint is shown after the joint was tested to failure.* The end fastener has sheared off, and it is visually apparent that the bolt deformation decreased rapidly from the end fastener toward the fasteners in the middle of the joint. An apparent double shear condition existed in the first six or seven fasteners of region 1, as indicated by the deformation along both shear planes. Thereafter, the fasteners resisted the load in single shear, transferring the load primarily to the lap plates adjacent to the main plate cutoffs. Although the fasteners in a symmetric butt joint are loaded in double shear, the fasteners in a shingle joint may be loaded either in single or double shear, depending on their location within the joint.

*In order to use the same bolt lot in all tests it was necessary (see Fig. 7.4) for the bolts in this particular joint to have less than full thread engagement for the nuts. Control tests indicated that the full bolt shear capacity was obtained even with less than full thread engagement. This practice is not recommended for field installations, however.

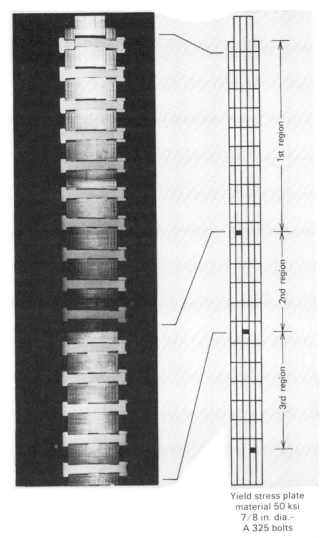

Yield stress plate
material 50 ksi
7/8 in. dia.–
A 325 bolts

Fig. 7.4. Sawed section of a three-region shingle joint after loading to failure.

Tests on riveted shingle joints showed an overall behavior that was comparable to the behavior of bolted shingle joints.[3.8,7.3] Riveted joints exhibited less slip than the bolted joints, because there is less hole clearance. When fastener failure is the governing failure mode, the overall deformation of large riveted shingle joints is likely to exceed the comparable deformation of an otherwise identical bolted joint.[7.2] This is primarily because of the different load versus deformation characteristics of rivets as compared with high-strength bolts.

7.3   JOINT STIFFNESS

The stiffness of a joint is characterized by the slope of its load versus deformation diagram. Figures 7.2 and 7.3 indicate that the total load is transferred initially by friction on the faying surfaces of the joint. It is also apparent that the stiffness of shingle joints is not significantly affected by a slip of the connection. Only yielding of the gross or net section causes a decrease in joint stiffness. Since the working load level does not exceed the yield strength of the net section, the joint stiffness may be considered to be reflected by the full cross-section, with an area equal to the total gross area of the main and lap plates. A comparable condition was observed with symmetric butt joints.

7.4   LOAD PARTITION AND ULTIMATE STRENGTH

The analytical solution for load partition and ultimate strength of shingle joints is based on a mathematical model that is similar to that used for symmetric butt joints as described earlier. The butt joint is a special case of a shingle joint.[7.2] The same basic assumptions which are discussed in Subsection 5.2.5 still apply. In addition, it is assumed that the transfer of load between the lap plates and the main plate takes place along the two planes that are common to the main plate core as illustrated in Fig. 7.5. Thus, no relative movement between the various plies of the lap plate or between the various plies of the main plate is considered. Each segment of the lap plate and main plate between consecutive fasteners is assumed to function as a unit with properties that are the aggregate of the constitutent plies. The model assumes the top and bottom lap plates to be a single plate of variable thickness, comparable to the main plate. This idealization results in regions of variable length with uniform plate properties within each region.

The force versus displacement relationships for plies of uniform width as well as for the fasteners, are those empirically developed in Ref. 5.22. The solution is comparable to the solution for a symmetrical butt splice.[7.2] The theoretical results were in good agreement with the experimental data on bolted shingle joints.[7.1] It was concluded that the load partition and ultimate strength can be predicted within acceptable limits if double shear behavior is assumed in the first region and single shear behavior in the interior regions of the shingle joints. This assumption is examined in greater detail in Section 7.5.

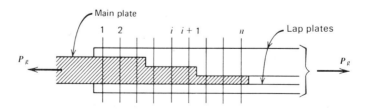

Fig. 7.5.   Idealized model of a shingle joint.

7.5 EFFECT OF JOINT GEOMETRY

The theoretical solution was used to study analytically the effects of various joint geometries on the ultimate strength.[7.1] The nondimensionalized ratio of the predicted ultimate strength to the working load of the joint, P_u/P_w, was used as an index of joint behavior. The working load was either based on the fastener shear area or on the net area of the main plate. Two possible assumptions for evaluating the total fastener shear in a joint were examined, namely (1) double shear of the fasteners throughout the joint, and (2) double shear in the first region and single shear in the other regions.

In the analytical study the yield stress and tensile strength of the plate material were assumed as 60 and 88 ksi, respectively, resulting in a 35 ksi allowable tensile stress for the plate material. The joints were fastened by $\frac{7}{8}$-in. dia. A325 bolts of minimum specified mechanical properties. The fastener pitch was held constant at 3 in.

The variables studies were (1) the A_n/A_s ratio, defined as the ratio of the net main plate area in the first region to the total effective fastener shear area; (2) the total number of fasteners in a joint; (3) the number of fasteners per region; and (4) the number of regions.

7.5.1 Effect of Variation in A_n/A_s Ratio and Joint Length

Figure 7.6 shows the change in joint strength with length for different A_n/A_s ratios ranging from 0.375 to 1.00 for shingle joints with three equal length regions. The fasteners were assumed to act in double shear in all three regions for one series of studies, and the results are indicated by the open dots. Each curve represents a different allowable shear stress. For example, an A_n/A_s ratio of 0.625 corresponds to an allowable shear stress of 22 ksi for double shear. Test results have indicated that the joint strength is likely to be overestimated for joints with high A_n/A_s ratios. This was primarily due to the single shear behavior observed in the interior regions.[7.1,7.3]

The analysis was also made assuming single shear behavior of the fasteners in the interior regions. These results are also shown in Fig. 7.6. It is apparent that for lower A_n/A_s ratios it does not matter whether double or single shear is assumed in the interior regions. For these joints the fasteners in the first region are the critical ones, as is illustrated in Fig. 7.7. At higher A_n/A_s levels, the load carried by the interior fasteners was greater, and a reduction in effective shear area had a more pronounced influence on joint strength (see Fig. 7.6). This was confirmed by the experimental results.[7.1]

7.5.2 Number of Fasteners per Region

The effect of varying the number of fasteners in each region was studied analytically by shifting an equal number of fasteners from each interior region into the first region. The total number of fasteners in the joint as well as the plate areas were not changed. Double shear behavior of the fasteners was assumed in the first region, and single shear behavior was assumed in the interior regions. The results are

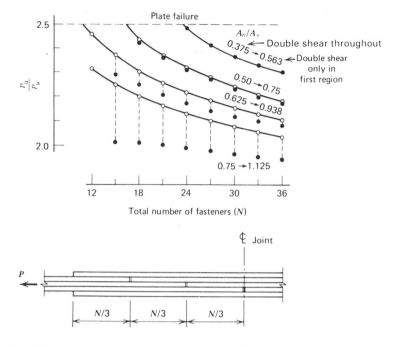

Fig. 7.6. Effect of assuming single shear in interior regions. ○ Analytical prediction assuming double shear in region 1; ● Analytical prediction assuming double shear. Single shear in interior regions.

summarized in Fig. 7.8. Sometimes a fastener failure was predicted in the interior regions when the fasteners were rearranged.[7.1] At the 0.75 A_n/A_s level, this only occurred in the short joints when four fasteners were shifted into the first region. No variation in strength occurred in the longer joints.

At the 1.125 A_n/A_s level, slight increases in strength were predicted by shifting fasteners into the first region.

From this study it was concluded that the predicted strength of shingle joints of a given length was not greatly influenced by rearranging the fasteners. This trend was also confirmed by the test data reported in Ref. 7.1.

7.5.3 Number of Regions

The effects of varying the number of main plate terminations was studied by comparing the strengths of joints with one, two, and three regions. All joints had the same total number of fasteners and the same plate areas. In the case of multiple region joints, an equal number of fasteners was provided per region. Double shear behavior of the fasteners was assumed in the first region, with single shear in the interior regions. The one-region joints were symmetrical butt joints having the total main plate area terminated at one location.

Figure 7.9 shows the change in ratio P_u/P_w due to the variation in the number

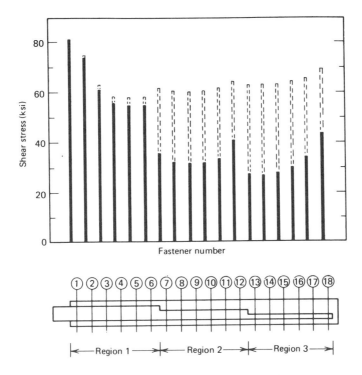

Fig. 7.7. Fastener shear distribution assuming single or double shear in interior regions. For $A_n/A_s =$ 0.50, double shear asssumed in all regions, $P_u = 930$ kips. For $A_n/A_s = 0.75$, double shear in first region, single shear in interior regions, $P_u = 927$ kips.

of regions. Note that the A_n/A_s ratio increases as the number of regions increases. This results from the assumed shear behavior of the fasteners in the interior regions. As indicated in Fig. 7.9, for the joints represented by the solid dots (A_n/A_s ratio is equal to 0.50 for the single region joint) there was no appreciable change in strength as the number of regions was changed. At the higher A_n/A_s ratios, indicated by the open dots in Fig. 7.9, the two- and three-region joints were less efficient. Greater variation was apparent for the shorter lengths. However, it is doubtful that short joints will be shingled.

At higher A_n/A_s ratios, the distribution of load to interior fasteners was greater than at lower A_n/A_s ratios. Thus, terminating the main plates at different locations and reducing the effective shear area resulted in a reduction in strength.

7.6 DESIGN RECOMMENDATIONS

7.6.1 Approximate Method of Analysis

Like other types of connections, shingle joints are statically indeterminant; thus, the distribution of forces depends on the relative deformations of the component

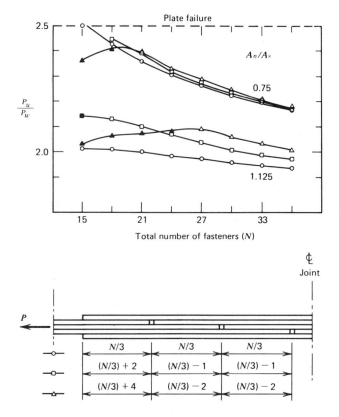

Fig. 7.8. Effect of rearranging fasteners. ■ ▲ Denote failure in interior regions.

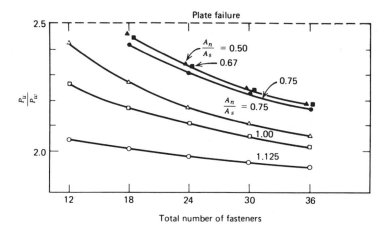

Fig. 7.9. Effect of number of regions. ▲ △ One-region joint (double shear is assumed). ■ □ Two-region joint: double shear in first region and single shear in interior regions. ● ○ Three-region joint; $A_n/A_s = 0.50$, open symbols for the single region joint; $A_n/A_s = 0.75$, solid symbols for the single region joint.

members and fasteners. The condition is further complicated in shingle joints by the unsymmetric positioning of main plate terminations. Analytical elastic solutions that predict the distribution of load in the main and splice plates of shingle joints have been developed.[7.5] The solution has been extended into the plastic range to predict the ultimate strength of the connection.[7.2] These theoretical analyses, however, are too cumbersome and impractical for ordinary design practice. Simplifying assumptions must be made that reduce the solution for design to one based primarily on equilibrium.

There are several existing methods for estimating the distribution of force in the main and lap plates of a shingle splice. Two of the most popular methods are:[7.4]

1. Forces in splice plates are inversely proportional to their distances from the member being spliced.
2. Forces in each member at a section through a splice are proportional to their areas.

In method 1, it is assumed that at each discontinuity the amount of force distributed to the lap plates is proportional to the area of the member being terminated. The forces in the continuous main members are assumed to remain unchanged. This is illustrated schematically in Fig. 7.10a. The transfer of load is made in the region directly preceding the point of termination, and it is assumed that the original load is restored to the spliced member in the region following termination.

In method 2 (see Fig. 7.10b), the total applied load is assumed to be distributed to all continuous members at the position of a main plate termination in proportion to their areas. No direct assumption is made regarding the amount of load transferred to the splice plates in a particular region as in method 1. If the lap plates are of equal area, method 2 predicts that the shear transfer is equal along the top and bottom shear planes in the first region, regardless of their positions with respect to the member being terminated.

Previous shingle joint tests have shown that at each plate discontinuity, there was a sudden pick-up of load in the adjacent plate elements.[3.8,7.5] Another approximate method of analysis was developed on the basis of these observations and test results. This method, referred to as method 3 and illustrated in Fig. 7.10c, assumes that the total load is distributed to all members at a section through the joint in proportion to their areas, first considering the terminated members as being continuous. The load assumed to be carried by a terminating member is then distributed to the two adjacent plates in proportion to their areas. Hence, a two-stage distribution is used.

Figure 7.11 compares the measured plate forces in a three-region test joint with the three design methods.[7.1] The partition of load was determined from the measured plate strains at different cross-sections along the length. The comparisons were at the working load levels as determined by the main plate net areas. It is apparent from Fig. 7.11 that method 1 underestimated the total transfer of load in the first and second region. Loads substantially greater than those estimated by

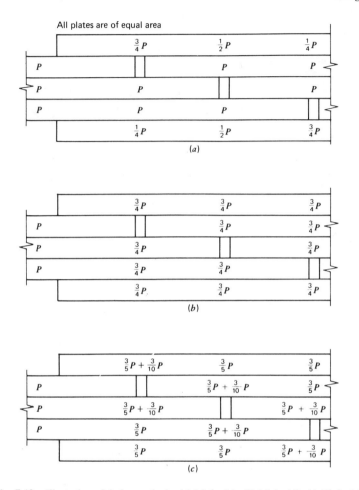

Fig. 7.10. Illustration of design methods. (*a*) Method 1. (*b*) Method 2. (*c*) Method 3.

method 1 were measured in the bottom lap plates. Test results indicated that the force in the top and bottom plates were nearly equal in the first region.

The distribution of load in the main plates of the joint as determined by method 2 was in good agreement with the measured forces. Slight variation between the theoretical distribution and test results occurred in the top and bottom lap plates. It was found that this method slightly underestimates the forces in the plates adjacent to a plate termination.

The distributions of force determined by method 3 provided the best correlation with the test results, as shown in Fig. 7.11. The method provided a reasonable estimate of the force distributions in all joint components and accurately predicts a more effective use of the fasteners in the interior regions, thus requiring less fasteners than the other methods. This method is therefore recommended for design purposes.

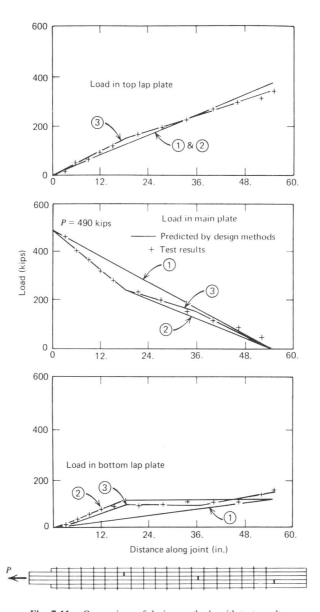

Fig. 7.11. Comparison of design methods with test results.

For design it is recommended that method 3 be used to approximate the load distribution in the plates and fasteners. With this method, it is also recommended that the first region of shingle splices have double lap plates of equal area. This reduces the critical shear transfer along the plate adjacent to the first plate termination.

Where practical, it is also recommended that the top and bottom lap plates have equal lengths in the first region. As shown in Fig. 7.4, equal deformation was observed along both shear planes at failure. It is believed that equal length splice plates would more effectively utilize the critical end fasteners.

With the introduction of a gusset into the splice as in a truss joint, however, additional fasteners are required along the shear plane adjacent to the gusset to transfer load from diagonal members. Since these fasteners are not required along the bottom shear plane, it is believed that the bottom lap plates can be shorter than the top lap plate in the first region if an adequate number of fasteners is still provided.

7.6.2 Connected Material

Once the load distribution throughout the plates is determined, the plate dimensions can be obtained. The design recommendations given in Subsections 5.4.3, and 5.4.4 for the connected plates are also applicable to shingle joints.

7.6.3 Fasteners

After the load partition has been established, the required number of fasteners per region can be determined. The difference in plate load between two adjacent plates is transmitted by shear of the fasteners. An examination of all possible shear planes in each region results in one or more critical shear planes for each region. The number of fasteners is readily determined from the shear resistance of the fasteners.

The design recommendations given in Subsection 5.4.2 for slip-resistant and other bolted joints subjected to static loading conditions are also applicable to the design of slip-resistant and other bolted shingle joints. The design shear stress for shingle joints depends on the bolt quality as well as on the joint length. Since the first region is the critical one in most shingle joints, the design shear stress for non-slip-critical shingle joints should be reduced by 20% if the length of the first region exceeds 50 in. All other design recommendations given in Subsection 5.4.2 are applicable to shingle joints.

REFERENCES

7.1 E. Power and J. W. Fisher, "Behavior and Design of Shingle Joints," *Journal of the Structural Division, ASCE*, Vol. 98, ST9, September 1972.

7.2 S. C. Desai and J. W. Fisher, *Analysis of Shingle Joints*, Fritz Laboratory Report 340.5, Lehigh University, Bethlehem, Pennsylvania, 1970.

7.3 E. Davis, G. B. Woodruff, and H. E. Davis, "Tension Tests of Large Riveted Joints," *Transactions, ASCE*, Vol. 66, No. 8, Part 2, pp. 1193–1299, 1940.

7.4 W. J. Yusavage (Ed.), *Simple Span Deck Truss Bridge*, Manual of Bridge Design Practice, 2nd ed., State of California, Highway Transportation Agency, Department of Public Works, Division of Highways, Sacramento, 1963.

7.5 U. Rivera and J. W. Fisher, *Load Partition and Ultimate Strength of Shingle Joints*, Fritz Laboratory Report 340.6, Lehigh University, Bethlehem, Pennsylvania, 1970.

Chapter Eight
Lap Joints

8.1 INTRODUCTION

In contrast to bolts in symmetric butt splices, fasteners in lap splices have only one shear plane. Depending on the geometry of the joint and the loading conditions, the behavior of lap joints may differ significantly from the behavior of symmetric butt joints with the fasteners loaded in double shear.

The simplest type of lap splice is shown in Fig. 8.1a. Such joints are simple to fabricate and erect but are usually avoided because of concern with the inherent eccentricity that results in deformations such as those shown in Fig. 8.1a. These effects of bending may be minimized by providing restraining diaphragms or stiffeners that restrict the rotation and out-of-plane displacement of the joint. Such restraints may be an integral part of the member. Often situations arise in which the restraints are provided by the connected member itself; a typical example is the hanger connection shown in Fig. 8.1b or the flange splices of a girder (Fig. 8.1c). Because of symmetry of the shearing planes and diaphragm action of the web, bending of the lap splice does not occur in any significant amount, although the fasteners are in a single shear condition and an eccentricity of the load exists.

Fasteners in a lap splice are mainly subjected to shear. However, depending on joint geometry and loading conditions, bending can result in an additional tensile component in the fastener. As noted in the following sections, this tensile component is often of minor importance and does not affect significantly the ultimate strength of the connection.

8.2 BEHAVIOR OF LAP JOINTS

In a discussion of the behavior of lap joints it is convenient to define two categories of lap joints as follows:

1. Joints in which restraints are provided so that bending can be neglected (Fig. 8.1b and c).

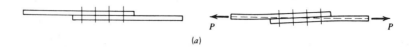

(a)

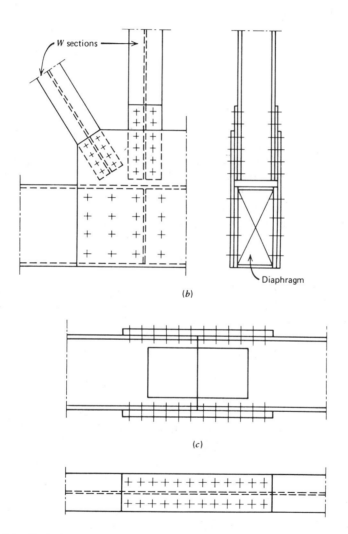

(b)

(c)

Fig. 8.1. Typical lap splices with fasteners subjected to single shear. (a) Lap splice connection. (b) Typical connection in truss-type bridge. (c) Girder splice.

2. Joints that are not restrained against bending. In these joints secondary bending stresses are developed due to the eccentricity of the load.

Static tension tests of lap joints with restraint against out-of-plane deformation exhibit a load versus deformation behavior that is essentially comparable to the behavior observed for symmetric butt joints (see Fig. 8.2). The slip resistance and the ultimate strength of single shear lap splices was found to equal one-half the double shear resistance provided by a butt joint. As expected, the "unbuttoning" behavior (as discussed in Chapter 5) was also observed in long lap joints.[4.6, 8.1]

The load versus deformation behavior of lap joints that were not restrained against out-of-plane displacement has been examined with small joints with two or three fasteners in a line.[6.2, 8.2, 8.3] Since restraints were not provided, the joints showed considerable deformation due to the eccentricity of the load, as shown in Fig. 8.3. It is evident that the effects of bending are mainly confined to the regions where plate discontinuities occur. Obviously, as the joint length increases, bending will become less pronounced, and the influence on the behavior of the connection should decrease. The influence of bending is most pronounced in a splice with only a single fastener in the direction of the applied load. In such a joint the fastener is not only subjected to single shear, but a secondary tensile component may be present as well. Furthermore, the plate material in the direct vicinity of the splice is subjected to high bending stresses due to the eccentricity of the load. However, this has little influence on the load capacity, since the material will strain-harden and cause yielding on the gross area of the connected plate.

Tests on single bolt lap splices showed that the slip resistance was not noticeably affected by the additional bending.[8.2, 8.3] Shear failures of the fasteners were observed at an average fastener shear stress that was about 10% less than observed in symmetric butt joints with similar material properties. Hence, the bending tended to decrease slightly the ultimate strength of short connections. The shear strength

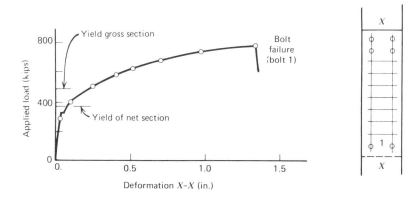

Fig. 8.2. Typical load versus deformation curve for lap joints in which restraints against bending are provided.

Fig. 8.3. Single shear specimen after test. (Courtesy of U.S. Steel Corp.)

of longer lap joints with no restraints against bending should not be as affected by the effects of bending.

Lap joints may be subjected to a repeated type loading as well. The critical joint component under such loading conditions is not the fastener but the plate material. A severe decrease in the plate fatigue strength is apparent in unrestrained lap joints when compared with butt joints.[6.2] The bending deformations cause larger stress ranges to occur at the discontinuities of the joint. The bending stress combines with the normal stress and results in high local stresses that reduce the fatigue strength. The reduction in fatigue strength depends on the joint geometry and the magnitude of the secondary bending. Hence, single shear splices subject to stress cycles should not be used unless the out-of-plane bending deformations are prevented.[6.2]

8.3 DESIGN RECOMMENDATIONS

When designing lap joints, both the fasteners and the plate material should be considered. Consideration should also be given to the type of loading and whether out-of-plane deformation will adversely affect the joint performance.

8.3.1 Static Loading Conditions

It was concluded earlier that the average shear strength of the fasteners at ultimate load and the slip resistance of lap joints are in reasonable agreement with the behavior observed on comparable symmetric butt joints. Therefore, the design recommendations given in Chapter 5 are applicable to lap joints for static type loading conditions. Bending of the joint does not significantly influence the slip resistance or strength. Hence, the provisions provided in Chapter 5 for both bolts and plate material are applicable.

8.3.2 Repeated-Type Loading

Since the plate is the critical element under repeated loads, lap joints should only be used under repeated loading conditions when secondary bending stresses are prevented or minimized. This requires suitable stiffening or joint geometry which will prevent out-of-plane movement. Lap connections that are susceptible to out-of-plane movements should not be used under repeated loading conditions. The design recommendations given in Chapter 5 for the plate material of symmetric butt joints are applicable as well to the design of lap joints that are not subjected to bending effects.

REFERENCES

8.1 R. A. Bendigo, J. W. Fisher, and J. L. Rumpf, *Static Tension Tests of Bolted Lap Joints*, Fritz Engineering Laboratory Report 271.9, Lehigh University, Bethlehem, Pennsylvania, August 1962.

8.2 Z. Shoukry and W. T. Haisch, "Bolted Connections with Varied Hole Diameters," *Journal of the Structural Division, ASCE*, Vol. 96, ST6, June 1970.

8.3 K. D. Ives, *Evaluation of Oversize Holes in Friction-Type Single Shear Joints*, Bulletin Applied Research Laboratory, U.S. Steel Corporation, Pittsburgh, Pennsylvania, June 1971.

Chapter Nine
Oversize and Slotted Holes

9.1 INTRODUCTION

Since the first application of high-strength bolts in 1947, bolt holes $\frac{1}{16}$ in. larger than the bolts have been used for assembly. A similar practice was adopted in Europe and Japan, where a hole diameter 2 mm greater than the nominal bolt diameter became standard practice.[9.1]

Restricting the nominal hole diameter to $\frac{1}{16}$ in. in excess of the nominal bolt diameter can impose rigid alignment conditions between structural members, particularly in large joints. Sometimes erection problems occur when the holes in the plate material do not line up properly because of mismatching. Occasionally, steel fabricators must preassemble structures to ensure that the joint will align properly during erection. With a larger hole size, it is possible to eliminate the preassembly process and save both time and money. To determine the feasibility of oversize holes, it was necessary to evaluate the performance of bolted connections with greater amounts of oversize.

An oversize hole provides the same clearance in all directions to meet tolerances during erection. However, if an adjustment is needed in a particular direction, slotted holes can be used, as shown in Fig. 9.1a and b. Slotted holes are identified by their parallel or transverse alignment with respect to the direction of the applied load (see Fig. 9.1a and b).

When oversize and slotted holes are used, additional plate material is removed from the vicinity of high clamping forces. The influence of this condition on the behavior of connections has been investigated experimentally.[4.26, 8.2, 8.3, 9.1, 9.3] The effect of oversize and slotted holes on such factors as the loss in bolt tension after installation, the slip resistance, and the ultimate strength of shear splices has been examined. Tightening procedures were studied as well. Provisions based on these findings are now included in specifications.[1.4]

9.2 EFFECT OF HOLE SIZE ON BOLT TENSION AND INSTALLATION

The load versus deformation characteristics of joints assembled with high-strength bolts installed in oversize or slotted holes depend, among other factors, on the bolt

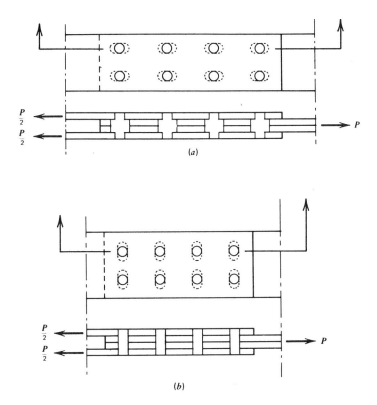

Fig. 9.1. Slotted holes. (*a*) Parallel slotted holes. (*b*) Transverse slotted holes.

clamping force. Hence, it is necessary to examine the effect of varying hole diameters on the bolt installation. This includes the degree of scouring around the hole and the clamping force induced by standard installation procedures. These factors are of primary interest when slip-resistant joints are used.

Tests have indicated that oversize and slotted holes can significantly influence the level of bolt preload when bolts are installed in accordance with common practice.[4.26] This is illustrated in Fig. 9.2, where the observed bolt tension after installation by the turn-of-the-nut method is shown for several different hole clearances.[4.26] The 1-in. dia. A325 bolts installed in $1\frac{1}{4}$-in. dia. holes, that is, with $\frac{1}{4}$-in. clearance, showed that the average bolt tension was about the same irrespective of whether or not a washer was used under the nut. The bolt tension attained was about 118% of the required minimum tension. This is about 15% lower than the average tension that is observed in joints with the normal $\frac{1}{16}$-in. clearance (Subsection 5.1.7). Depressions in the plate occurred under the bolt heads during tightening and were greater than the depressions observed with the usual $\frac{1}{16}$-in. hole clearance. Severe galling of both plate and nut occurred with oversize holes when washers were omitted from under the turned element, as is illustrated in Figs. 9.3 and 9.4.[4.26] One-inch diameter bolts installed with only one washer under the turned

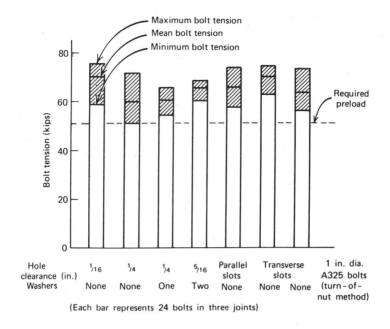

Fig. 9.2. Range of bolt tensions for normal, oversize, and slotted holes.

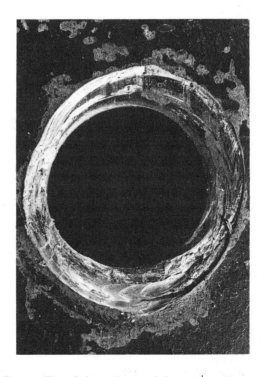

Fig. 9.3. Severe galling of plate under turned element ($\frac{1}{4}$ in. clearance, no washer).

Fig. 9.4. Plate area under element in which washer was used ($\frac{1}{4}$ in. clearance).

element in $1\frac{5}{16}$-in. diameter holes (not shown in Fig. 9.2) failed to achieve their minimum required tension. The bolt heads had recessed severely into the plate around the holes. When washers were placed under both the nut and bolt head, the range of bolt tension achieved ranged from 110 to 144% of the minimum required tension, with an average value of 125%. In other, unpublished, tests, large diameter ($1\frac{1}{8}$-in.) A490 bolts were installed in $\frac{5}{16}$-in. oversize holes. Standard washers were used under both the nut and the bolt head. Although scouring was observed, it was principally dishing of the washers under the very high preload that prevented the specified minimum preload from being attained. Only when thicker washers were used ($\frac{5}{16}$ in.) could the specified minimum preload be obtained in these tests.

The depression of the bolt into the plate or the dishing of the washer means that prescribed rotation of the nut may not produce the required amount of bolt elongation. Consequently, the bolt preload may be less than that specified. In the calibrated wrench procedure, if the deformation characteristic of the calibrator is stiffer than that of the joint with oversize holes, the same problem can arise.

Assuming that the bearing pressure developed under the flat areas of the bolt heads with $\frac{1}{4}$-in. clearance holes is the maximum permitted on A36 steel plate, a theoretical maximum hole clearance for any size bolt can be determined. The area of the plate remaining under the flat of the bolt head must be sufficient so that this

Table 9.1. Hole Clearance for Different Hole Sizes

Bolt Size	Maximum Hole Diameter (in.)	Amount of Clearance
$\frac{1}{2}$	$\frac{11}{16}$	$\frac{3}{16}$
$\frac{5}{8}$	$\frac{13}{16}$	$\frac{3}{16}$
$\frac{3}{4}$	$\frac{15}{16}$	$\frac{3}{16}$
$\frac{7}{8}$	$1\frac{1}{16}$	$\frac{3}{16}$
1	$1\frac{1}{4}$	$\frac{1}{4}$
$1\frac{1}{8}$	$1\frac{7}{16}$	$\frac{5}{16}$
$1\frac{1}{4}$	$1\frac{9}{16}$	$\frac{5}{16}$
$1\frac{3}{4}$	$1\frac{11}{16}$	$\frac{5}{16}$
$1\frac{1}{2}$	$1\frac{13}{16}$	$\frac{5}{16}$

pressure is not exceeded. The results of such computations are summarized in Table 9.1. The hole diameters have been rounded off to the nearest sixteenth of an inch. All of the available test results substantiate that the specified minimum preload can be reached or exceeded for A325 bolts if the hole and bolt diameter combinations shown in Table 9.1 are used. As has already been noted, additional precautions in the form of thicker washers will be necessary for large diameter A490 bolts. Bolts installed by the turn-of-nut method in slotted holes also showed a decrease in the mean bolt tension when compared with similar bolts installed in standard holes with a $\frac{1}{16}$ in. oversize.[4.26] Hence, the use of either oversize or slotted holes is likely to reduce slightly the mean clamping force in the fastener.

Immediately after a bolt is tightened, a loss in bolt tension occurs. This is thought to result from creep and plastic deformation in the threaded portions and plastic flow in the steel plates under the head and the nut. These deformations result in an elastic recovery and subsequent loss in bolt tension. Studies on bolts installed in holes with a standard hole clearance are summarized in Ref. 4.26 and in Chapter 4. In general, the total loss in preload was about 5 to 10% of the initial preload, depending on grip length (3 to 6 in.) and whether washers were used. Most of the loss in preload occurred within a short time after the bolt was tightened.

A few relaxation tests have been conducted on bolts installed in oversize holes and are reported in Ref. 4.26. It was observed that none of the variations in the hole diameter or the presence of slots had any significant effect on this loss. Virtually all of the losses occurred within 1 week after installation, as was also observed with earlier studies. The loss in tension was observed to be about 8% of the initial preload. This is directly comparable to earlier studies on regular size holes with a standard clearance of $\frac{1}{16}$ in.

9.3 JOINT BEHAVIOR

9.3.1 Slip Resistance

Figure 9.5 shows typical load versus slip relationships of joints with oversize or slotted holes.[4.26] The response is almost linear until the load approaches the major

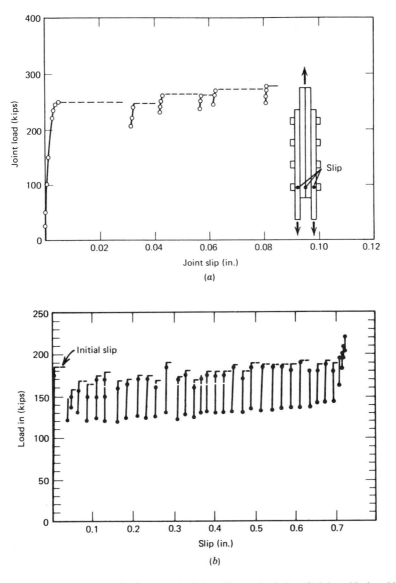

Fig. 9.5. Typical load versus slip diagrams. (*a*) Joint with oversize holes. (*b*) Joint with slotted holes.

slip load. The initial slip was always observed to be less than the amount of hole clearance. Subsequent loading of the joint after major slip had occurred produced small slips until the joint came into bearing. These small slips occurred at loads near the major slip load. The test results shown in Fig. 9.5 were obtained using double shear splices like those illustrated in Fig. 9.1[4.26] The fasteners were 1-in. dia. A325 bolts, and the connected material was A36 steel in the clean mill scale condition. A summary of the observed slip coefficients as a function of the hole geometry for both oversize and slotted hole conditions is shown in Fig. 9.6. It was

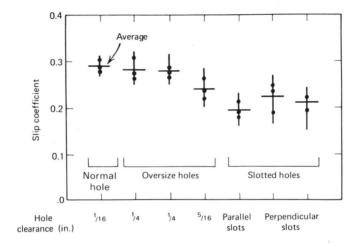

Fig. 9.6. Comparison of average slip coefficients.

concluded that the average slip coefficient for joints with up to $\frac{1}{4}$-in. hole clearance did not change with varying oversize. The joints with $\frac{5}{16}$-in. clearance holes showed a 17% decrease in the slip coefficient for clean mill scale faying surfaces. The slip coefficient for joints with slotted holes showed a 22 to 33% decrease when compared with test specimens with a hole clearance of $\frac{1}{16}$ in. A decrease in slip resistance with the removal of plate material from around the bolt was expected because of the resulting high contact pressures in the area around the bolt. Removal of the plate causes extremely high contact pressures adjacent to the bolt holes that tends to flatten the surface irregularities and thereby reduces the slip resistance of the joint.

The slip resistance is also affected by the decreased clamping force that has been observed in joints with oversize and slotted holes. The combined effects of the change in slip coefficient and the reduction in the clamping force on the slip resistance is estimated to cause a 15% reduction in slip resistance for oversize holes and a 30% reduction for parallel and transverse slotted holes.[4.26]

Major slip of the connection is terminated when one or more bolts come into bearing against the plates. The amount of slip exhibited before bearing occurs depends on the available clearance and fabrication tolerances. Joints with oversize holes or parallel slotted holes may undergo substantial displacements if the slip resistance of the joint is exceeded.

Studies have also been carried out to evaluate the influence of oversize holes upon the slip resistance of blast-cleaned and coated surfaces.[9.3] This work showed that, for holes up to $\frac{1}{4}$-in. greater in diameter than the bolt diameter, there was no significant effect of hole oversize on the slip coefficient. (Further work with sand-blasted surfaces showed that the surface roughness of the A572 steel surfaces did not significantly affect the slip coefficient, and that sandblasting time did not affect the slip coefficient for A36, A572, and A514 steels tested. These tests were carried out using joints with holes of normal clearance.)

The painted surfaces examined included organic zinc primer, with or without an epoxy topcoat, and inorganic zinc primer with a vinyl topcoat. The specified primer thickness was 6 mils and that of the topcoat was 3 mils. This part of the study again found that holes up to $\frac{1}{4}$ in. greater in diameter than the $\frac{7}{8}$-in. diameter bolts did not affect the slip resistance of the joints.

Although joints with slotted holes were not examined in this study, it is reasonable to expect that their slip behavior would be similar to that displayed by the coated or blast-cleaned surfaces containing oversize holes.

9.3.2 Ultimate Strength

The ultimate strength of a connection is governed by either the shear capacity of the bolts or the tensile capacity of the plates. The effect of oversize holes or slotted holes on the ultimate strength can be evaluated by examination of the limiting case, transverse slotted holes. Tests have shown that the presence of transverse slotted holes does not result in a reduction of the tensile strength of the plates or of the shear strength of the fasteners.[4.26] Hence, the ultimate strength of a joint can be assumed to be unaffected by either oversize or slotted holes.

9.4 DESIGN RECOMMENDATIONS

Since the ultimate strength of a joint with oversize or slotted holes is the same as the ultimate strength of a similar standard type connection with identical bolt and plate areas, the design recommendations given in Chapter 5 are applicable. The provisions given there for both plate material and bolts of bearing-type shear splices are applicable also to joints with oversize or slotted holes. Care must be exercised when using oversize or slotted holes to ensure that excessive deformation will not occur at working loads. The slots should be oriented so that large displacements cannot result. Transverse slotted holes are preferable, since they limit the slip to the same magnitude that can be experienced with standard hole clearances.

Design recommendations for slip-resistant joints with oversize or slotted holes must reflect the reduced slip resistance. Hole diameters that do not exceed those given in Table 9.1 do not significantly alter the slip coefficient. However, the clamping force is reduced by about 15%, and this must be reflected in the slip resistance and design conditions. A factor 0.85 can be used to provide for the reduced clamping force and its effect on the slip resistance. For slip-resistant joints with slotted holes, a reduction factor of 0.70 will account for the loss in slip resistance caused by either parallel or slotted holes.

To prevent the use of extremely large slotted holes, present specifications limit the length of slotted holes to $2\frac{1}{2}$ times the bolt diameter. (These are defined as long slotted holes.) The width of the hole should not exceed the bolt diameter by more than $\frac{1}{16}$ in. Short slotted holes are also used. Short slotted holes are $\frac{1}{16}$ in. wider than the bolt diameter and have a length that does not exceed the allowable oversize diameter for that bolt size by more than $\frac{1}{16}$ in. Joints with short slotted holes will develop the same slip resistance as joints with oversize holes. Therefore, the design of joints with oversized or short slotted holes is the same.

To achieve an adequate clamping force in the bolts, washers should be used under both the bolt head and the nut when oversize or slotted holes occur in the outside plates of a joint. Special requirements are necessary for large diameter A490 bolts.

DESIGN RECOMMENDATIONS FOR OVERSIZE AND SLOTTED HOLES

Hardened washers are to be inserted under both the head and the nut if oversize or slotted holes are placed in the outside plies of a connection. A490 bolts with diameters greater than 1 in. should have at least $\frac{5}{16}$-in. thickness material under both the head and the nut in order to bridge over a slotted or oversize hole. (Use of multiple washers to make up the thickness will not be satisfactory.) If this additional material is hardened, no washers will be necessary. However, if ordinary structural steel plate is used, standard hardened washers should be added under both the nut and bolt head.

Slip-Resistant Joints

$P'_s = 0.85 \, P_s$ for oversize and short slotted holes not exceeding the dimensions given in Table 9.2

$P'_s = 0.70 \, P_s$ for long slotted holes not exceeding the dimensions given in Table 9.2

where P_s is the slip load described in Subsection 5.4.2 for joints using holes of normal clearance.

For coated surfaces, the design recommendations given in Section 12.5 should be similarly modified if slotted or oversize holes are present.

Table 9.2 Standard, Oversize, and Slotted Hole Dimensions

Bolt Diam	Hole Dimensions			
	Standard (Diam)	Oversize (Diam)	Short Slot (Width × Length)	Long Slot (Width × Length)
$\frac{1}{2}$	$\frac{9}{16}$	$\frac{5}{8}$	$\frac{9}{16} \times \frac{11}{16}$	$\frac{9}{16} \times 1\frac{1}{4}$
$\frac{5}{8}$	$\frac{11}{16}$	$\frac{13}{16}$	$\frac{11}{16} \times \frac{7}{8}$	$\frac{11}{16} \times 1\frac{9}{16}$
$\frac{3}{4}$	$\frac{13}{16}$	$\frac{15}{16}$	$\frac{13}{16} \times 1$	$\frac{13}{16} \times 1\frac{7}{8}$
$\frac{7}{8}$	$\frac{15}{16}$	$1\frac{1}{16}$	$\frac{15}{16} \times 1\frac{1}{8}$	$\frac{15}{16} \times 2\frac{3}{16}$
1	$1\frac{1}{16}$	$1\frac{1}{4}$	$1\frac{1}{16} \times 1\frac{5}{16}$	$1\frac{1}{16} \times 2\frac{1}{2}$
$\geq 1\frac{1}{8}$	$d+\frac{1}{16}$	$d+\frac{5}{16}$	$(d+\frac{1}{16}) \times (d+\frac{3}{8})$	$(d+\frac{1}{16}) \times (2.5 \times d)$

REFERENCES

9.1 European Convention for Constructional Steelwork, *Specifications for Assembly of Structural Joints Using High Strength Bolts*, 3rd ed., Rotterdam, The Netherlands, April 1971.

9.2 O. Steinhardt, K. Möhler, and G. Valtinat, *Versuche zur Anwendung Vorgespannter Schrauben im Stahlbau, Teil IV*, Bericht des Deutschen Auschusses für Stahlbau, Stahlbau-Verlag Gmbh, Cologne, Germany, February 1969.

9.3 K.H. Frank and J.A. Yura, *An Experimental Study of Bolted Shear Connections*, Report No. FHWA/RD-81/148, U.S. Department of Transportation, Washington, D.C., December 1981.

Chapter Ten

Filler Plates Between Surfaces

10.1 INTRODUCTION

Often splices are symmetric and consist of identical structural components on each side of the splice. The joint components share a number of common shear planes, and splice plates are used to transfer the load across the splice. In other cases, however, it may be necessary to connect members of different dimensions, or gaps may be intentionally created in order to provide for easier erection. In these cases, the joint must be filled out in its thickness dimensions so that there are common faying surfaces and shear planes on each side of the joint and there are no significant joint eccentricities. This packing is accomplished by means of filler plates. The beam or girder splice with different depth members on each side of the joint, as illustrated in Fig. 10.1, is a typical example of a joint using filler plates. Filler plates are also frequently encountered in splices of axially loaded built-up members in truss bridges.

The influence of filler plates on the load transfer through a splice comprising one or more filler plates is discussed in this chapter. There are not a great deal of experimental data available, but tests have been carried out to determine both the slip resistance and the ultimate strength of bolted joints in which fillers are present. A series of tests was carried out in England in 1965 on single bolt joints with $\frac{1}{8}$-in. thick washers inserted between faying surfaces.[10.1] Tests were also reported by Lee and Fisher on four bolt joints with blast-cleaned surfaces and fillers.[5.10] The filler thickness varied from $\frac{1}{16}$ to 1 in. Yura et al.[10.2] have reported on tests that used both two and three bolts in line, fillers of various thicknesses, both tight and loose fillers, and the use of multiple plies as compared with single thickness fillers. Their work provided information on both the slip behavior and the ultimate strength of the joints. Although the available data are rather limited, they provide a reasonable indication of the behavior of joints with filler plates.

10.2 TYPES OF FILLER PLATES AND LOAD TRANSFER

Filler plates are classified as ''loose'' or ''tight'' fillers. In the case of loose fillers, the plates are solely used as packing pieces. Their only function is to provide a

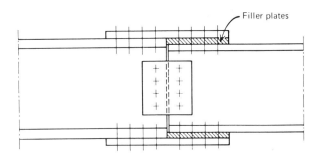

Fig. 10.1. Beam or girder splice with filler plates.

common shear plane on each side of the splice, as shown in Fig. 10.2a. Tight fillers are also used as packing pieces, but the fillers are extended beyond the splice plates and the joint is made longer. As with loose fillers, tight fillers also function to provide a common shear plane on each side of the joint. However, as shown in Fig. 10.2b, tight fillers are connected by additional fasteners outside the main splice, and they become an integral part of the connection. Tight fillers are said to be "developed" if they extend far enough beyond the main splice so that a uniform stress pattern occurs through both the connected material and the filler plate.

In slip-resistant joints, the load is transferred by frictional forces acting on the contact surfaces. Hence, the fasteners are not loaded in direct shear, as they are in a bearing-type joint. Therefore, loose fillers are adequate for slip-resistant joints when the surface condition of the joint components provides adequate slip resistance, and the forces can all be transferred on the faying surfaces. Test results reported in Refs. 10.1 and 5.10 support this conclusion. The tests reported in Ref.

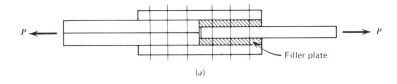

(a)

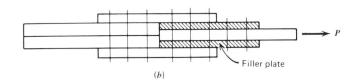

(b)

Fig. 10.2. Types of filler plates. (a) Loose fillers. (b) Tight fillers.

10.1 are summarized in Fig. 10.3. All specimens had two bolts in line, packed with $\frac{1}{8}$-in. thick washers of variable diameter in order to control the contact area. It is readily apparent from Fig. 10.3 that the insertion of $\frac{1}{8}$-in. thick "loose" fillers between the joint faying surfaces did not significantly affect the slip resistance. This was observed to be true for both clean mill scale and blast-cleaned faying surfaces.

The tests reported by Lee and Fisher were on four bolt joints with blast-cleaned surfaces.[5.10] The fillers were symmetrically placed on both faying surfaces and varied in thickness from $\frac{1}{16}$ to 1 in. Figure 10.4 shows the joint arrangement as well as some typical test results. There seems to be no significant variation in the slip resistance with different thicknesses of the fillers. Furthermore, as shown in Fig. 10.5, the observed slip coefficients varied between 0.47 and 0.57, which is well within the 95% confidence limits for blast-cleaned surfaces summarized in Table 5.1.

In the slip tests done by Yura et al.[10.2], a specimen that had two bolts on one side of the splice location and three on the other was used. All faying surfaces were clean mill scale, but the filler material was A36 steel whereas all other joint

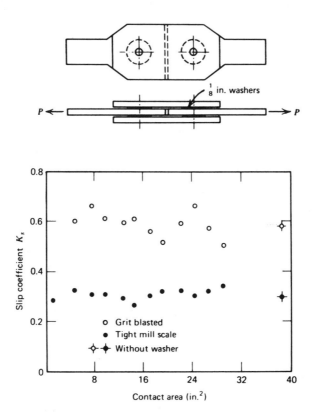

Fig. 10.3. Slip coefficient–contact area relationship for tests by Dorman Long and Company.

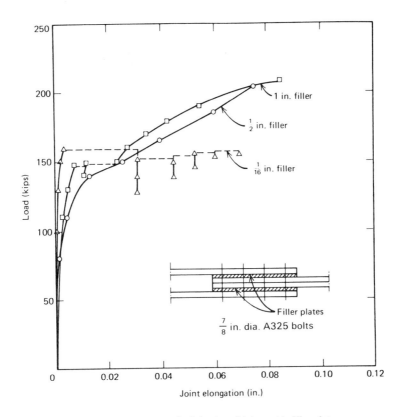

250

200   □
                                              ○
                                                       1 in. filler

                                          ○
                                    □   ½ in. filler

150 △   △   △ △ ▽ △
     □ □   △ ▽   ▽
                   ○ △   1/16 in. filler

     ▽
Load (kips)

100 □
     ○

 50   ┌─────────────┐
                             │   │
                             └─Filler plates┘
                              7/8 in. dia. A325 bolts

  0   0.02   0.04   0.06   0.08   0.10   0.12
                    Joint elongation (in.)

Fig. 10.4. Load versus slip behavior of joints with filler plates.

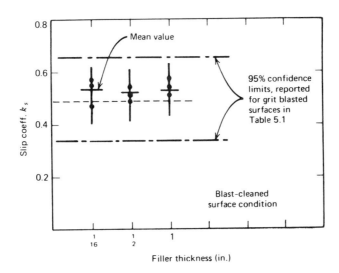

0.8

              Mean value

0.6   95% confidence
                                             limits, reported
                                             for grit blasted
                                             surfaces in
0.4   Table 5.1

Slip coeff. k_s

0.2
                           Blast-cleaned
                           surface condition

        1/16   1/2   1
              Filler thickness (in.)

Fig. 10.5. Comparison of slip coefficients.

and splice components were of A514 steel. Two filler plate conditions were used. In one case, a single $\frac{1}{4}$-in. filler was used under each splice plate, and in the other, three $\frac{1}{4}$-in. fillers were used under each splice plate. On the two-bolt side of the joint the fillers were loose, whereas on the three-bolt side they were developed by the addition of one additional fastener placed beyond the main splice plate. A control specimen that used no filler plates was also part of the program. Two specimens were tested for each of the configurations described.

There are no slip coefficient data for A514 steel in the clean mill scale condition outside the two tests done by Yura *et al.* Since these produced an average slip coefficient of 0.33, it can be assumed that these surfaces acted like those of lower grade steels in the clean mill scale condition (see Table 5.1). When one filler plate of $\frac{1}{4}$-in. thick A36 steel was inserted, a mean slip coefficient of 0.27 was obtained, and when three $\frac{1}{4}$-in. plies of A36 steel were used, the slip coefficient was 0.18. These results may be said to demonstrate a decreasing slip resistance with filler plate thickness and, possibly, an effect with respect to the number of plies used. It is interesting to note that no slips were recorded on the three-bolt side of these joints, even though that side was subjected to exactly the same load as the two-bolt side. For these tests, the three-bolt side never slipped, even up to the shear failure load of the two bolt side.

There is a conflict between the results of Lee and Fisher[5.10] and Yura *et al.*[10.2] The former showed that filler plate thicknesses up to 1 in. had no effect on the slip load, whereas the latter showed a decreased slip coefficient when one $\frac{1}{4}$-in. thick ply was used and an even larger decrease when three $\frac{1}{4}$-in. plies were used. In assessing the test results of Yura, *et al.*, it must be noted that all "slips" recorded were extremely small. For example, the movement at the slip load in one of the specimens with three $\frac{1}{4}$-in. plies was only about 0.02 in. (0.5 mm). This is approximately one-third of the nominal hole clearance, and such a small amount of movement would not be a cause for concern unless load reversal were present. Furthermore, all of the Yura *et al.* test results except one were within two standard deviations of the mean slip coefficient for clean mill scale surfaces, a confidence limit of 95%.

Based on these limited data, it is concluded that filler plates with a surface condition similar to that of the other components of the joint do not significantly affect the slip resistance of a bolted joint.

Vasarhelyi and Chen tested bolted butt joints with slightly different thickness main plates on each side of the joint.[10.3] Filler plates were not used, and consequently full surface contact could not be obtained adjacent to the end of the thinner main plate. Generally, a decrease in slip resistance was observed when compared with the control joints with main plates of equal thickness. They suggested that the slip resistance could be improved by increasing the distance from the plate edge to the first row of bolts. This would provide more flexibility in the lap plates and allow more clamping force to be used effectively for load transfer.

There is no doubt that the presence of loose filler plates has the potential for reducing the bolt shear strength. Fig. 10.6 shows the idealized loading of a bolt

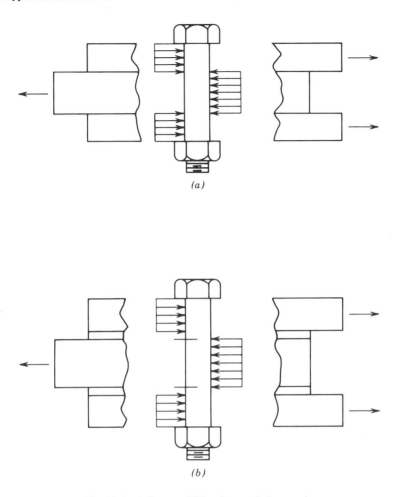

Fig. 10.6. Influence of filler plates on bolt strength.

in a shear splice after the slip load has been exceeded. No filler plates are present in the joint shown in Fig. 10.6a. The location of the potential shear planes is well-defined, and this constitutes the standard shear strength case described in Chapter 5. Fig. 10.6b shows the idealized loading for a bolt contained in a joint that uses loose filler plates. The location of the potential shear planes is no longer clear, and it is obvious that bolt bending can occur. The extent of the bending and its influence on the bolt shear strength will depend on the thickness of the filler plates. So far as is known, the tests conducted by Yura et al.[10.2] are the only ones to have explored this behavior. In addition to the specimens for which the slip resistance was obtained, specimens using a single 0.075-in. thick filler plate and a single $\frac{3}{4}$-in. thick filler plate were also tested.

When no fillers were present, the two-bolt side of the joints failed at 1.003 times the shear strength of single bolts taken from the same lot. This ratio was 0.974, 0.991, 0.877, and 0.863 for the cases of a single 0.075-in. filler, a single $\frac{1}{4}$-in. filler, a single $\frac{3}{4}$-in. filler, and three $\frac{1}{4}$-in. fillers, respectively. (Two joints were tested in each category, and the average results are quoted herein. There was close agreement between pairs of test results.) A reduction in bolt shear strength capacity is apparent for the larger filler plate thicknesses. The amount of bolt bending will be affected by the amount of bearing deformation in the plates immediately adjacent to the holes. (This deformation can be seen in Fig. 5.33.) Because these tests used A514 steel plates for the connected material, the amount of bearing deformation can be expected to be somewhat less than that which would occur in joints using lower yield strength steels. Thus, the shear strength reductions determined in the Yura *et al.* tests might represent minimum values.

The shear strength reduction with increasing filler plate thickness that was observed in the Yura *et al.* study must be the result of tensile forces in the bolts. These tensile forces will be the consequence of lap plate prying and bolt bending. Counteracting this effect, the shear area available increases as bending occurs because the shear plane no longer passes through the bolt shank at right angles to the longitudinal axis of the bolt. This phenomenon has been observed in many tests and is evident in the photographs of failed specimens in the Yura *et al.* study. Evidently, in these tests the shear strength reduction due to the presence of tensile forces in the bolt exceeded the shear strength increase present due to increased shear area.

Tight fillers might be advantageous or necessary if the bearing stress on the main plate rather than the shear capacity of the fastener governs the design. Providing a tight filler increases the thickness of the plate to be spliced and thereby reduces the bearing stress. There are no bolted joint tests with tight fillers available. However, tests have been conducted on riveted joints to verify the assumed behavior.

10.3 DESIGN RECOMMENDATIONS

Depending on the required load transfer, loose or tight fillers can be used in slip-resistant or bearing-type joints. For slip-resistant joints, loose fillers with surface conditions comparable to other joint components are capable of developing the required slip resistance. Slip-resistant joints do not require additional fasteners when filler plates are used. The fillers become integral components of the joint, and filler thickness does not significantly affect the joint behavior.

For bearing-type joints, where the load is transmitted by shear and bearing of the bolts, loose fillers can be used as long as excessive bending of the bolts does not occur. It is suggested that single loose fillers up to $\frac{1}{4}$-in. thick can be used without considering a reduction in bolt shear strength. If the loose filler thickness exceeds this, the bolt shear strength capacity should be reduced. A reduction of 15% would be appropriate for a loose filler thickness of $\frac{3}{4}$ in.

Tight fillers are not required in bearing-type joints if the allowable bearing stress on the main plate is not exceeded. Tests on riveted joints have indicated that tight fillers are desirable when thick filler plates are needed and long grips result. This requires additional fasteners and they are preferably placed outside the connection, as shown in Fig. 10.2b. As an alternative solution, the additional fasteners may be placed in the main splice.

The design recommendations given in Chapter 5 for the plates and fasteners are applicable to the design of connections with filler plates.

REFERENCES

10.1 L. G. Johnson, *High Strength Friction Grip Bolts*, unpublished report, Dorman Long and Company, England, September 1965.

10.2 J. A. Yura, M. A. Hansen, and K. H. Frank, "Bolted Splice Connections with Undeveloped Fillers," *Journal of the Structural Division, ASCE*, Vol. 108, ST12, December 1982.

10.3 D. D. Vasarhelyi and C. C. Chen, "Bolted Joints with Plates of Different Thickness," *Journal of the Structural Division, ASCE*, Vol. 93, ST6, December 1967.

Chapter Eleven
Alignment of Holes

11.1 INTRODUCTION

Holes in mechanically fastened joints are either punched, subpunched and reamed, or drilled, and the hole diameter is generally $\frac{1}{16}$ in. greater than the nominal bolt diameter. Since connections contain two or more fasteners, the alignment of holes is of concern. Usual shop practice is to fabricate the constituent parts of a joint separately. Since dimensional tolerances are necessary during the fabrication process, the holes of component parts of a joint are not likely to be perfectly aligned unless all plies are clamped together before drilling. Misalignment may also result from erection tolerances. Hence, it is desirable to ascertain whether hole offsets have detrimental effects on the joint behavior.

This chapter discusses the influence of misalignments on the behavior of high-strength bolted connections.

11.2 BEHAVIOR OF JOINTS WITH MISALIGNED HOLES

The experimental data available on joints with misaligned holes are not extensive. Vasarhelyi *et al.* have reported on a series of tests where misalignment was purposely introduced into the joint by providing mismatching holes.[11.1, 11.2]

The two major concerns with misaligned holes are whether the slip resistance is affected and whether the misalignment adversely affects the joint strength and performance. With joints transferring load by shear and bearing of the fasteners, bolts placed in misaligned holes obviously will come into bearing prior to other fasteners in the joint. If the fasteners and plates have sufficient ductility and can accommodate the unequal forces and displacements, the misalignments should not have a significant effect.

In addition to affecting the distribution of forces on the fasteners, misalignment may also influence the stress distribution in the connected plates of the joint.

Depending on the amount of misalignment in the hole pattern, tests on misaligned joints have indicated that slip generally develops more gradually as compared with joints with good alignment.[11.1, 11.2] This is expected, since full hole clearance slip is prevented due to the misalignment of the holes. As slip develops, the plates

194

come into bearing and the fasteners generally offer further resistance to the slip movement.

A series of small slips have been observed to develop at load levels considerably above the normal slip resistance.[11.1,11.2] These partial slips bring more bolts into bearing and result in geometric self-adjustment of the joint elements as the applied loads force alignment of the joint. The joint tends to pivot around fasteners already in bearing, and eventually this results in more bolts in bearing.

Tests have indicated that the slip resistance of a misaligned bolted joint is equal to or exceeds the slip resistance of a joint without misalignment. This is visually apparent in Fig. 11.1. As the misaligned condition was made more severe, there was not as much rigid body motion possible. No significant change in joint stiffness was apparent until the applied loads were nearly twice as large as the load that caused major slip to develop with good alignment. Comparable results have been observed with more complex joints where misalignment is more probable.[3.8,4.6] Misaligned holes always result in less movement between the connected plies. The joint stiffness is improved, and full hole slip is not possible.

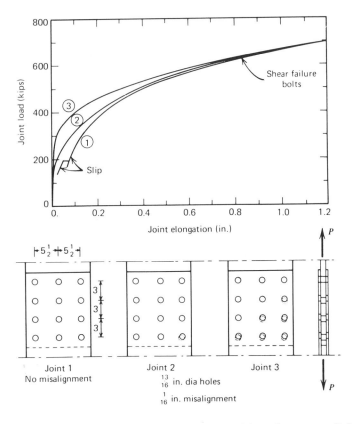

Fig. 11.1.   Influence of misalignment of holes on load versus deformation response (Ref. 11.2).

When slip develops, one or more bolts come into bearing. As the applied load is increased, these bolts and the adjacent plate material must deform so that other bolts can come into bearing as well. If the deformation capacity of the plates and the bolts will permit it, all bolts may come into bearing before shear failure develops in one or more bolts. Excessive misalignment may prevent all bolts from coming into bearing and prevent the full shear strength of the joint from being developed. This situation is somewhat analogous to the load partition that occurs in long bolted joints. The critical fastener may be subjected to severe deformations and fail prematurely before the full joint strength can be attained.[4.6]

The tests on compact bolted joints with different degrees of misalignment throughout the bolt pattern that are summarized in Fig. 11.1 show that misalignment has a negligible effect on the ultimate strength of the joints. If anything, the misalignment had a beneficial effect. It improved the slip resistance, decreased the rigid body motion between connected plies, offered a stiffer joint, and did not result in a decrease in joint strength. Comparable results were reported in later tests.[11.2]

As the connected material increases in yield and tensile strength, misalignment may have a more adverse effect. Not as much ductility is available for the redistribution of the load, and a critical fastener could be sheared off prematurely. This condition is also more critical with higher strength bolts, since they have less deformation capacity in shear. The plastic deformation capacity of the plate material and the deformation capacity of the bolt both contribute to the adjustment that occurs in the joint. Obviously, the more deformation capacity that is available, the better the redistribution of plate and bolt forces.

11.3 DESIGN RECOMMENDATIONS

The amount of misalignment in a joint depends largely on the joint geometry as well as on fabrication tolerances and erection procedures. Since bolt holes are generally $\frac{1}{16}$ in. in excess of the nominal bolt diameter, some adjustment possibility is provided. Available test results do not indicate any adverse effect of misalignment resulting from hole clearance on either the slip resistance or the ultimate strength of the joint.[11.1, 11.2] Hence, the usual misalignment that may result from erection or fabrication tolerances does not affect the design of joints.

Since the deformation capacity of the fasteners and plate material are of prime importance in the readjustment capacity of bolted joints with misaligned holes, the degree of tolerance will decrease when higher strength materials with lower ductility are used.

REFERENCES

11.1 D. D. Vasarhelyi, S. Y. Beano, R. B. Madison, Z. A. Lu, and U. C. Vasishth, "Effects of Fabrication Techniques on Bolted Joints," *Journal of the Structural Division, ASCE*, Vol. 85, ST3, March 1959.

11.2 D. D. Vasarhelyi and W. N. Chang, "Misalignment in Bolted Joints," *Journal of the Structural Division, ASCE*, Vol. 91, ST4, August 1965.

Chapter Twelve
Surface Coatings

12.1 INTRODUCTION

Situations often arise in steel construction in which it is desirable to provide a protective coating on the members and the faying surfaces of their joints. The treatment prevents corrosion due to exposure before erection or provides a corrosion-resistant layer to reduce maintenance costs during the lifetime of the structure. When the treatment is applied to prevent long term corrosion, the coating is of a permanent nature; usually, metallic layers of zinc or aluminum are employed. For temporary protective purposes, a wash primer is often used that is usually removed upon assembly by grinding or by dissolving with various solvents. Other less permanent coatings such as vinyl washes and linseed oil are also used.

It has long been recognized that protective coatings alter the slip characteristics of bolted joints to varying degrees.[4.18, 12.7] Consequently, the design of slip-resistant joints with coated faying surfaces must reflect the influence of such treatments on the slip resistance.

For bearing-type joints, the permissible load for both working stress design and load factor design is based on the ultimate strength of the connection. This strength is, of course, independent of any coating that may be used. Therefore, the comments in this chapter are confined to the influence of protective coatings on the response characteristics and performance of slip-resistant joints subjected to various types of loading.

In the past, the main means of providing corrosion protection to exposed surfaces was by galvanizing. Galvanized members were used mainly for transmission line towers, and the joints were often designed on the basis of their bolt shear strength so that bearing-type connections resulted.

In some structures, ribbed bearing bolts were used to minimize joint slip. At that time, the use of coatings for slip-resistant joints was limited or prohibited by the specifications.[1.4, 5.30] These restrictions were the result of early research that indicated that a low frictional resistance resulted when galvanized surfaces were present.[12.7] However, as a result of continuing research, protective surface treatments that provide adequate slip resistance have been developed.[4.11, 4.18, 4.27, 5.11, 5.17, 5.37, 9.1, 12.1–12.3] These studies indicate that adequate frictional resistance of coated surfaces can be achieved, and that coated high-strength bolts, nuts, and washers can be used provided that a suitable lubricant is used on the threaded part of the fastener (see Section 4.6). As a result of these studies, provisions were included in the RCRBSJ specification in 1970 that permitted certain surface treatments to be used in slip-resistant joints.[1.4] These treatments and their influence upon the load versus slip performance of slip-resistant joints are discussed in this chapter.

12.2 EFFECT OF TYPE OF COATING ON SHORT-DURATION SLIP RESISTANCE

When only temporary protection of the faying surface is needed, paints are often placed on the weather-exposed surfaces. Vinyl-washes and linseed oil have also been used as substitutes for red lead and similar paints.[5.11] If a more permanent protective coating is required, a metallic layer with a high corrosion resistance must be applied to the structural element. The most commonly used protective coatings can be classified as follows:

1. Hot-dip galvanizing, with or without a preassembly treatment, to improve the slip resistance of the surface
2. Metallizing with either sprayed zinc, aluminum, or a combination of both metals
3. Zinc-rich paints that use organic or inorganic vehicles
4. Vinyl washes or paints.

The effects of these coatings on the slip resistance of connections subjected to short-duration, statically applied loads are discussed in this section. Other factors, such as the load versus deformation behavior under sustained or repeated loading conditions, must be considered when applicable; they are discussed in subsequent sections.

12.2.1 Hot-Dip Galvanizing

The hot-dip galvanizing process requires the removal of the mill scale prior to the coating application. This is usually done by pickling the member in a bath of acid. Subsequently, the member is coated with a metallic layer by dipping into a bath of hot metal. Iron–zinc alloys or pure zinc are generally used for this process.

Test results indicate that hot-dip galvanizing generally results in a low frictional resistance of the faying surfaces.[4.11,4.18,12.1,12.13] Tests on joints with hot-dip galvanized faying surfaces have yielded slip coefficients between 0.08 and 0.36, with an average value of 0.19 (see Table 12.1).[12.13] The low slip resistance of galvanized surfaces as compared with clean mill scale surfaces is caused by the presence of the softer zinc layer that tends to act as a lubricant between the faying surfaces. Test results have also indicated that the slip coefficient decreases with an increase in coating thickness.[4.18,12.1]

Table 12.1. Slip Coefficients for Hot-Dip Galvanized Surfaces under Short-Duration Static Load

Ref.	Type of Treatment	Coating Thickness (mils)	Number of Tests	Average	Standard Deviation
4.18	Pickling in acid bath, hot-dip galvanized	2.4–5.0	10	0.23	0.023
12.1	Pickling in acid bath hot-dip galvanized	4.0	3	0.15	—
12.13	Pickling in acid bath, hot-dip galvanized (tests performed on one-bolt compression-type specimens)	—	15	0.21	0.08
12.1	Pickling in acid bath, hot-dip galvanized.	—	2	0.15	—
12.5	Pickling in acid bath, hot-dip galvanized	3.2	—	0.20	—
	Sand-blasted, hot-dip galvanized	3.2	—	0.28	—
12.13	Summary Study (data from various sources)	—	95	0.19	Value Min. 0.08 Max. 0.36 Estimated standard deviation 0.045

Note: 1 mil = 0.001 in. or 25.4 μm; a zinc coating of 1 oz/ft^2 corresponds to a coating thickness of 0.0017 in.

Variability in thickness of the metallic layer is inherent with the galvanizing process. Different treatment methods have also contributed to the variability observed for different test series. These factors are believed to be the major reasons for the relatively large scatter in the test data.[12.13]

The influence of the treatment method on the slip resistance of galvanized joints is illustrated by the test data summarized in Table 12.2. In these test series, all joint components were grit-blasted before pickling and subsequent dipping into the metal bath. Dipping time, cooling rate, and bath temperature were varied. For the plain, uncoated, blast-cleaned surfaces, an average slip coefficient of 0.73 resulted. The galvanized surfaces yielded average slip coefficients between 0.27 and 0.57.[12.8] The study indicated that the type of coating process can affect the slip resistance of the coated surfaces. These results, as well as data reported in Ref. 5.17, show that blast cleaning the surface before hot-dip galvanizing results in an improvement

Table 12.2.   Influence of Pregalvanizing Treatment on Slip Coefficient

Conditions	Series A	Series B	Series C
Surface condition	Grit-blasted to white metal	Hot-dip galvanized	Hot-dip galvanized
Coating thickness (mils)	—	4.0	4.5
Coating structure	—	Fe-Zn alloys 40%; pure zinc 60%	Fe-Zn alloys 100%
Preparation	—	Grit blasting and successive pickling with HCl	Grit blasting and successive pickling with HCl
Zinc bath temperature	—	452°C	467°C
Dipping time	—	1 min	3 min
Cooling	—	High: air blowing within 20 sec after withdrawal and successive water quenching	Low: specimens kept over the bath surface for 3 min, successive water quenching

Ref.	Type of Treatment	Coating Thickness (mils)	Number of Tests	Average	Standard Deviation
12.8	Grit-blasted, series A	—	10	0.73	0.05
	Grit-blasted, hot-dip galvanized, series C	4.2	10	0.57	0.01
	Grit-blasted, hot-dip galvanized, series B	4.2	10	0.27	0.03
5.17	Grit-blasted, hot-dip galvanized	4.0	12	0.30	—

of the slip resistance.[5.17, 12.5, 12.8] This results from the increased surface roughness due to the blast cleaning.

A significant improvement in the slip resistance of galvanized surfaces can be achieved by preassembly treatment of the contact surfaces. Among the treatments examined are wire brushing, sand or grit blasting, and a chemical treatment of the galvanized surfaces.[4.11] Wire brushing can be accomplished manually or with a power brush. A light blast cleaning that dulls the normal shiny appearance of the galvanized coating is generally sufficient. With either treatment it is essential to visibly alter the surface condition. However, in order to maintain the corrosion protection for which the galvanizing was applied in the first place, it is important

Table 12.3. Summary of Slip Coefficients of Hot-Dip Galvanized Surfaces (Determined from Compression-Type Specimens)

Surface Treatment	Average	Standard Deviation	Number of Tests
As-received	0.21	0.08	12
Weathered	0.20	0.06	17
Wire-brushed	0.37	0.01	6
Sand-blasted	0.44	0.02	9
Shot-blasted	0.37	0.10	6
Acetone-cleaned	0.32	0.03	9
Phosphate-treated	0.38	0.03	10
Chromate-treated	0.26	0.02	6

that the continuity of the coating not be disrupted. A substantial increase in slip resistance has been observed for some of the treatments.[4.11,4.27,12.13]

Slip coefficients obtained using small compression jigs are summarized in Table 12.3. Tests on larger tension connections with the same surface treatments have yielded somewhat lower slip coefficients. The results of the compression shear jig tests clearly show that an improvement in slip resistance can be obtained by wire brushing or lightly blast cleaning the galvanized surfaces of the joints prior to assembly. Tests reported in Ref. 5.17 yielded the same trend. Hence, treatment of hot-dip galvanized surfaces can result in a slip coefficient that is at least comparable to the coefficient for clean mill scale surfaces (see Fig. 12.1). Further tests are desirable to provide a better estimate of the slip coefficient for such surface conditions.

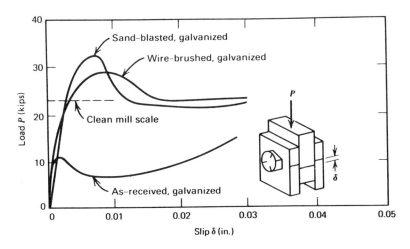

Fig. 12.1. Small shear block specimen tests indicate several surface treatments that enhance the frictional resistance of galvanized steel.

In contrast to clean mill scale or blast-cleaned surface conditions, a sudden slip does not usually occur in hot-dip galvanized joints. The observed slip is often gradual, with loads increasing until the bolts come into bearing.

12.2.2 Metallizing

The metallizing process involves spraying a hot metal onto the surfaces of a structural element to provide corrosion resistance. Zinc and aluminum are commonly used for metallizing structural members.

The surface to be metallized should have all oil and grease removed and must be roughened by blasting. The sprayed metal will only bond adequately to cleaned and roughened surfaces. Sand, crushed slag, or chilled iron grit are commonly used for blast cleaning the surface. The coating is applied shortly after blast cleaning. Different spraying processes can be used, and detail procedures are given in Ref. 12.11.

Short-duration slip tests on metallized surfaces have shown that high slip resistance can be achieved with these treatments.[4.18,5.17,5.37,12.1,12.5,12.6] Test results from metallized joints with various coating thicknesses are summarized in Table 12.4. It is apparent that the slip coefficient is related to the coating thickness. When the coating is thick compared with the surface irregularities resulting from blast cleaning, a relatively low slip coefficient results. Very thin coatings, 0.0005 to 0.001 in. (15 to 25 μm), also result in relatively low slip coefficients. The optimum slip performance was achieved when the coating thickness was between 0.002 and 0.004 in. (50 and 100 μm). The test data also indicate a higher slip coefficient for aluminum-sprayed surfaces as compared with zinc-sprayed surfaces with the same coating thickness. This difference in behavior is believed due to the difference in hardness of the metallic layer. A higher slip coefficient results with the harder aluminum coating.

Sealing treatments are often used to improve the corrosion resistance of the surfaces and to enhance their appearance.[12.11] These additional treatments tend to fill the surface irregularities and provide a smoother faying surface. This results in a decreased frictional resistance and a lower slip coefficient. Hence, sealing treatment should not be used on slip-resistant joints.[12.2]

12.2.3 Zinc-Rich Paints

Zinc-rich paints are coatings that contain a high zinc dust content; they provide a hard, abrasion-resistant protection for the coated surfaces.[12.12] They are mainly used for permanent or long-term corrosion protection. Some of the coatings are used for prefabrication or shop primers. The primer coats do not require as great a thickness as coatings for long-term protection.

Zinc-rich paints are available in a large number of different commercial mixes. These coatings use either organic or inorganic vehicles. Among the organic vehicles used are vinyls, epoxies, and polyesters.[5.11,12.12] Common inorganic vehicles are silicates, phosphates, and modifications thereof. Many of these coatings are supplied with the zinc-rich pigment packaged separately, and the materials are mixed

Table 12.4. Slip Coefficients for Metallized Surfaces (Short-Duration Tension-Type Tests)

Ref.	Type of Treatment	Coating Thickness (mils)	Number of Tests	Average	Standard Deviation
12.6	Corundum blast-cleaned,	0.8–1.6	—	0.42	—
	zinc-sprayed	2.0–2.8	—	0.45	—
		3.6–4.4	—	0.41	—
5.37	Sand-blasted,	8.0	—	0.40	—
	zinc-sprayed				
4.18	Sand-blasted,	—	2	0.48	—
	zinc-sprayed				
12.1	Grit-blasted,	3.0	—	0.78	—
	zinc-sprayed	0.6–1.0	20	0.42	0.04
	Shot-blasted,	—	—	0.60	—
	zinc-sprayed	3.0	—	0.70	—
	Corundum-blasted,	0.6–1.0	10	0.43	0.04
	zinc-sprayed				
12.5	Sand-blasted,	1.6	17	0.70	0.05
	zinc-sprayed				
	Sand-blasted	2.6	14	0.73	0.05
	two layers,				
	zinc-sprayed				
5.17	Grit-blasted,	4.0	12	0.82	—
	zinc-sprayed (*compression* type specimens)				
12.6	Corundum-blasted, aluminum-sprayed	0.8–1.6	—	0.56	—
		2.0–2.8	—	0.58	—
		3.6–4.4	—	0.59	—
12.1	Shot-blasted,	—	—	0.64	—
	aluminum-sprayed	4.0	—	0.79	—
	Grit-blasted,	1.6–2.2	20	0.74	0.08
	aluminum-sprayed	4.0	—	0.76	—
	Corundum-blasted,	1.6–2.2	10	0.73	0.10
	aluminum-sprayed				
5.37	Sand-blasted,	10.0	—	0.4	—
	aluminum-sprayed				
12.5	Sand-blasted,	2.4	—	0.67	—
	aluminum-sprayed				
12.6	Corundum-blasted, zinc-sprayed, aluminum-sprayed,	Layer thickness			
		Zn:1.2	—	0.49	—
		Al:1.2			
		Zn:1.2	—	0.42	—
		Al:4.0			

Table 12.4. (*Continued*)

Ref.	Type of Treatment	Coating Thickness (mils)	Number of Tests	Average	Standard Deviation
5.37	Sand blasted, Chrome-nickel sprayed	20.0	—	0.41	—
12.5	Sand-blasted, stainless steel sprayed	1.6	6	0.72	0.05

at the time of application. Depending on the chemical composition, these coatings may have a pot life as low as 6 hr.

The inorganic coatings are very resistant to solvents and oil and are also resistant to high humidity. The weathering resistance of inorganic coatings is outstanding because the coating continues to cure during prolonged exposure.[12.12] For best results, the inorganic coatings should be used over blast-cleaned surfaces that provide a "near-white" condition.

Compared with the inorganic coatings, organic coatings are generally more tolerant to variations in surface preparation. They tend to be more flexible but are also less tough and abrasion resistant than the inorganic materials.[12.12]

The slip behavior of connections with contact surfaces treated with zinc-rich paints that use inorganic or organic vehicles has been examined by tests.[9.2, 12.1] Table 12.5 summarizes the results of tests in which the faying surfaces were treated with zinc-rich paint using organic vehicles.[9.2, 12.1, 12.14] Prior to painting, the surfaces were cleaned either by grit blasting or sand blasting. The various paints tested included one- and two-component zinc dust paints, special primers, organic zinc primer, and organic zinc primer with an epoxy top coat. In one series of tests, uncoated blast-cleaned or sand-blasted surfaces were used as a control.

Considering grit-blasted and coated joints, a trend toward increasing slip resistance with increasing coating thickness is noted (up to 1.8 mils). However, the increase is not large. Bare, grit-blasted steel in these tests had a slip coefficient of 0.56, whereas grit-blasted surfaces with one-component zinc dust paint or special primer had a mean slip coefficient of 0.44. Use of a two-component zinc dust paint gave a lower slip coefficient, 0.31.

Uncoated, sand-blasted joints had a slip coefficient of 0.52. Addition of a primer lowered this to 0.20. Coating with special primer resulted in a slip coefficient of 0.41. Use of zinc dust paint gave slip coefficients that were dependent upon coating thickness, 0.8 mils giving a slip coefficient of 0.39, and 1.2 mils giving 0.23. (These two results were obtained from different test programs, however.) Sand-blasted joints with an organic zinc primer gave a slip coefficient of 0.46, and with an epoxy top coat the slip coefficient was 0.27.

In summary, the results tabulated in Table 12.5 show that any zinc-rich coating

Table 12.5. Slip Coefficient of Zinc-Rich Painted Surfaces (Organic Vehicles)

Ref.	Type of Treatment	Coating Thickness (mils) Primer Coat	Top Coat	Number of Tests	Average	Standard Deviation
9.2	Grit-blasted	—	—	6	0.56	0.01
	Grit-blasted, one		0.6	6	0.40	0.02
	component zinc		1.2	6	0.45	0.01
	dust paint		1.8	6	0.46	0.01
	Grit-blasted, spe-	0.6		6	0.39	0.01
	cial primer	1.2		6	0.41	0.02
		1.8		6	0.42	0.02
	Grit-blasted, two		0.6	6	0.30	0.02
	component zinc		1.2	6	0.31	0.03
	dust paint		1.8	6	0.33	0.01
	Sand blasted	—	—	10	0.52	0.04
	Sand-blasted, primer	0.6–0.8		8	0.20	0.02
	Sand-blasted, spe- cial primer	0.8		10	0.41	0.02
	Sand blasted, zinc dust paint		0.8	10	0.39	0.02
12.1	Sand-blasted, zinc dust paint		1.2	10	0.23	0.03
12.14	Sand-blasted, or- ganic zinc primer	2.5–10.3		94	0.47	0.07
	Sand-blasted, or- ganic zinc primer	2.8–9.7	1.8–4.0	90	0.27	0.03

applied to a grit-blasted or sand-blasted steel surface results in a lowered slip coefficient. In the majority of treatments listed, the slip coefficient was still comparable to that for sand or shot-blasted hot-dip galvanized surfaces, however. Still, it must be recognized that the treatment of steel surfaces with zinc-rich paint results in a wide variation in slip coefficient, depending on the treatment used.

Use of inorganic zinc-rich paints for coating provides better slip resistance than when organic zinc-rich paints are used.[9.2, 12.1, 12.6, 12.14] When zinc silicate paint has been used with a clear lacquer (water-glass) as a binding agent and zinc dust powder as the pigment, high slip resistance has resulted. The increased hardness of the zinc silicate coating provides a more slip-resistant surface than surfaces treated with organic zinc-rich paints. For optimum results, these paints are generally applied to blast-cleaned surfaces by either spraying or brushing.[12.12]

The thickness of zinc silicate coatings also slightly influences the slip coefficient. This is illustrated in Table 12.6, where test results for different coating thickness are summarized.[12.6] The specimens were all blast cleaned and then coated with zinc silicate paint supplied by five different suppliers. An increase in coating thickness increased the slip resistance for all five mixes.

Tests were performed in Germany on sand-blasted specimens treated with zinc silicate paint,[9.2] and the results are given in Table 12.7. The zinc paint was provided by five different suppliers, and the coating thickness varied from 1.0 to 1.4 mils. All five coatings provided slip coefficients that were about the same as plain sand-blasted surfaces. The maximum difference in average slip coefficients between coated and uncoated specimens was only about 13%. The other test results shown in Table 12.7 also show slip coefficients comparable to uncoated, sand-blasted, or grit-blasted surfaces.

The results also indicate that the chemical composition of the paint does not greatly influence the slip behavior. Much greater variation was observed with organic zinc-rich paints. Blast-cleaned surfaces treated with zinc silicate paints are likely to yield a slip coefficient that is about the same as the slip coefficient provided by blast-cleaned base metal.

12.2.4 Vinyl-Treated Surfaces

Vinyl washes or vinyl paints are also used for the corrosion protection of faying surfaces or bolted joints. They are easily applied, give moderate to good corrosion protection, and are relatively inexpensive. In the earlier studies reported,[5.9,5.11] two different vinyl washes were evaluated. These were applied in light coats (0.3 to 0.5 mils thick) on surfaces that had been sand blasted. Joints were tested both relatively soon after preparation and assembly and after 2 months exposure (prior to assembly) in an industrial atmosphere. Table 12.8 provides a summary of the test results. These studies showed that the exposure time prior to assembly did not have the significant influence on the slip coefficient. The slip coefficient for these joints in the sand-blasted and vinyl wash condition is not greatly reduced as compared with uncoated, clean mill scale conditions. (It should be noted that the

Table 12.6. Slip Coefficients for Surfaces Treated with Zinc Silicate Paint[a]

Coating Thickness (mils)	Average Slip Coefficient Product Identification				
	A	B	C	D	E
0.8	0.41	0.47	0.62	0.53	0.50
2.0	0.52	0.53	0.64	0.56	0.52

[a] Specimens were blast cleaned and treated with different zinc silicate paints. These results are averaged from two readings each (see Ref. 12.6).

Table 12.7.   Slip Coefficient of Zinc-Rich Painted Surfaces (Inorganic Vehicles)

Ref.	Type of Treatment	Coating Thickness (mils)		Number of Tests	Average	Standard Deviation
		Primer Coat	Top Coat			
9.2	Sand-blasted	—	—	4	0.61	—
	Sand-blasted, zinc silicate paint		1.4	6	0.61	0.01
			1.2	6	0.59	0.02
			1.2	6	0.52	0.01
			1.0	6	0.53	0.01
			1.0	6	0.58	0.04
			1.0	6	0.60	0.01
12.1	Grit-blasted, zinc silicate paint		0.8–1.6	10	0.57	0.07
			1.6–3.8	10	0.68	0.05
	Shot-blasted	—	—	—	0.60	—
	Shot-blasted zinc silicate paint		3.0	—	0.63	—
	Grit-blasted	—	—	—	0.58	—
	Grit-blasted, zinc silicate paint		3.0	—	0.56	—
12.14	Sand-blasted, inorganic zinc primer, vinyl top coat	2.1–11.5	1.3–2.8	80	0.50	0.06
	Sand-blasted, inorganic zinc primer (80% zinc)	6.8–8.0		10	0.61	0.03
	Sand-blasted, inorganic zinc primer (75% zinc)	6.7–7.2		5	0.51	0.01

specification governing use of the vinyl wash requires that the surfaces be sand blasted prior to application of the coating.)

A more recent study has examined the effect of using thicker vinyl coatings, presumably required for superior corrosion protection.[12.14] A summary of these results is also given in Table 12.8. In one series, a vinyl primer treatment 2.1 to 2.7 mils thick was applied to a blast-cleaned surface, and in another both a vinyl primer (2.3 to 2.8 mils) and a vinyl topcoat (2.1 to 2.5 mils) were used. The results

Table 12.8. Slip Coefficients of Vinyl-Treated Surfaces

Ref.	Type of Treatment	Coating Thickness (mils)		Number of Tests	Average	Standard Deviation
		Primer Coat	Top Coat			
5.9	Sand-blasted, vinyl wash (MIL-P15328A)	0.3–0.5		3	0.28	0.01
	Sand-blasted, vinyl wash (MIL-C15328A)	0.3–0.5		3	0.27	0.01
5.11	Sand-blasted, vinyl wash (MIL-C15328A)	0.3–0.5		3	0.29	0.02
	Sand-blasted, vinyl wash (MIL-C15328A), exposed 2 months	0.3–0.5		3	0.27	0.05
	Sand-blasted, vinyl wash (MIL-P15328B), exposed 2 months	0.3–0.5		3	0.26	0.01
12.14	Sand-blasted, vinyl primer	2.1–2.7		15	0.19	0.02
	Sand-blasted, vinyl primer, vinyl top coat	2.3–2.8	2.1–2.5	6	0.20	0.01

from these two series of tests were almost identical, and they show that a significant decrease in slip coefficient occurs as compared with the vinyl washes.

12.3 JOINT BEHAVIOR UNDER SUSTAINED LOADING

Early field experience and test results indicated that some galvanized members had a tendency to continue to slip under sustained loading.[12.3, 12.6, 12.9] Slip was only stopped when the bolts came into bearing. In some situations this small slippage may impair the serviceability of the structure. Hence, if a joint is subjected to sustained loading conditions and is slip critical, the slip performance of the coating, whether galvanized or treated otherwise, must be considered under the sustained load condition.

Laboratory tests have been performed to evaluate the load versus deformation behavior of different types of coated surfaces subjected to sustained loading.[12.3, 12.6, 12.9, 12.15] In general, the observed slip behavior with respect to time can be characterized by one of the three relationships shown in Fig. 12.2.[12.9] Category

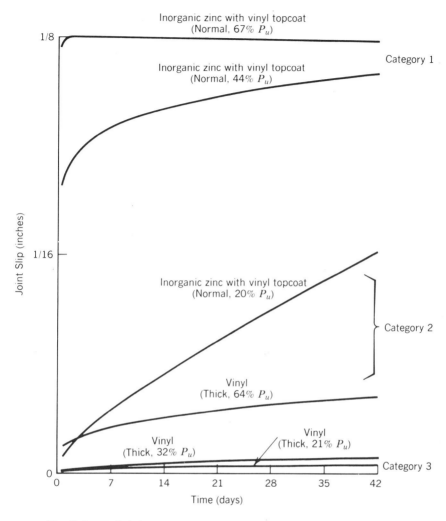

Fig. 12.2. Typical time versus slip curves for connections under sustained loading.

1 represents a class of connections in which major slip occurs during application of the load. The bolts come into bearing against the plate, and the joint remains stable with time unless the load reverses direction. Category 2 represents connections that do not initially slip into bearing but continue to slip under sustained loading; the connection is said to "creep". The slip rate under sustained loading only becomes zero when the bolts come into bearing. Category 3 shows joints with good slip resistance under both short term as well as sustained loading conditions. After a small initial extension, often elastic, no further slip is detected.

Test specimens are usually subjected to stepwise increasing load when evaluating slip resistance under sustained loading. After slip has been arrested or the slip rate has stabilized, the load is increased. This process is repeated until either the bolts are in bearing or the slip resistance of the faying surfaces is exceeded.

Tests on hot-dip galvanized joints and on joints with zinc-rich paints with vinyl topcoat tested at medium or high loads subjected to sustained loading show a steady-state rate of slip.[12.6, 12.9] The connections developed a creep-type behavior as indicated in Fig. 12.2. Preassembly treatments that yielded an increase in short-duration slip resistance did not significantly improve the slip behavior of the hot-dip galvanized joints under sustained loading. Joints treated with organic zinc-rich paints showed essentially the same behavior.[9.1] The zinc layer created by the zinc-rich paint acts like a lubricant between the surfaces, and this results in creep under sustained loading.

Better results were obtained using zinc silicate paint on the joint faying surfaces. Both short-duration slip resistance and sustained load slip resistance were improved. Test results indicated that a coating layer thickness equal to 0.0020 to 0.0024 in. (50 to 60 μm) provided about the same slip coefficient for sustained loading and short duration tests.[12.3, 12.6] Even when the sustained loads were close to the slip load of the connection, a stable joint condition resulted. Joints protected with only a vinyl coating and tested at medium to low sustained load levels also exhibited good performance.

Metallizing with either zinc or aluminum resulted in good short-duration slip resistance. However, under sustained loading conditions, aluminum-sprayed faying surfaces provided better slip resistance than zinc-sprayed surfaces. Slip coefficients for sprayed aluminum surfaces were found to be about the same for both the sustained and short-duration loading tests. Zinc-sprayed surfaces exhibited creep when the joint was subjected to loads that were close to the slip resistance of the surfaces. If an appropriate margin was applied so that the loads were well below the slip resistance of the joint, satisfactory sustained load characteristics were observed.[12.3, 12.6]

12.4 JOINT BEHAVIOR UNDER REPEATED LOADING

The behavior of plain, noncoated bolted butt joints subjected to repeated loading conditions is summarized in Chapter 5. For slip-resistant joints, crack initiation and growth were generally observed to occur through the gross section. When the slip resistance was decreased, failure usually occurred at the net section. The application of a protective surface coating may alter the slip resistance of a joint; therefore, its influence on the fatigue strength of a joint has to be examined.

Fatigue tests have been performed on hot-dip galvanized joints because they exhibited low slip resistance during short-duration static slip tests.[4.11, 4.27, 12.10, 12.13] The tests showed that the connection either slipped into bearing or the connection "locked up" (ceased to slip) after a few cycles when there was load reversal.[4.11, 12.10, 12.13] This locking-up effect is illustrated in Fig. 12.3 for a hot-dip galvanized joint subjected to repeated load reversal. Figure 12.3 shows that the displacements during the fifth cycle were about the same as those that occurred

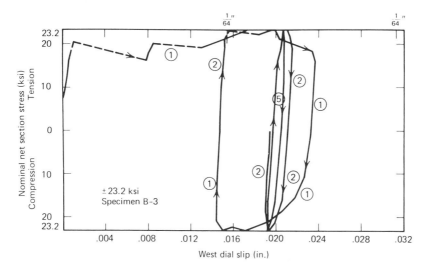

Fig. 12.3. "Lock-up" effect of hot-dip galvanized joints.

during the second cycle. Hence, small slips in hot-dip galvanized joints did not decrease the fatigue life. Failures often occurred through the gross section areas despite the initial slip.

Disassembly of these joints confirmed the tendency to lock up. To separate the plates of a joint it was often necessary to pry them apart.[4.11] Layers of zinc tended to pull off from the surfaces of the plate as a result of galling and seizing of the zinc coating in the region around the bolt holes where high contact pressures exist.

A preassembly treatment of the hot-dip galvanized surface by wire brushing or light blast cleaning did not influence the fatigue life.

The effect of repeated loading on joints with other protective surface treatments such as metallizing and zinc-rich paints has been studied as well.[5.17, 12.4, 12.6] The results of these tests have also indicated that the lower slip resistance and early slips in the joints did not influence the fatigue resistance of coated joints. Their fatigue strength was equal to or greater than the fatigue resistance of uncoated joints of similar dimensions.

In an attempt to explain this behavior qualitatively, joints were classified into two categories depending on whether or not the slip resistance of the test joint was exceeded by the applied load. It is shown in Chapter 5 that uncoated slip-resistant bolted joints subjected to repeated loading exhibit a fretting-type crack initiation in the gross section ahead of the bolt hole. Hence, a surface coating that provides sufficient slip resistance should provide comparable behavior. Such behavior was frequently observed in tests on metallized and zinc silicate painted joints.[9.2, 12.4] For these surface conditions, repeated cyclic loads close to the slip load of the connection did not result in significant slip in the connection.

12.5 DESIGN RECOMMENDATIONS

Joints with protective coatings should be designed by the criteria suggested in Chapter 5. Depending on the type of surface treatment, a wide range of slip coefficients is possible. Even for a specific type of treatment, substantial scatter can result from fabrication procedures.

Subsection 5.4.2 set out the design recommendations for slip-resistant connections in which the faying surfaces are uncoated. Alternative forms for expressing the slip load were given, one setting forth the slip load as a function of the specified minimum clamping force in the bolt ($T_{i_{\text{spec}}}$) and the mean slip coefficient of the faying surface ($k_{s_{\text{mean}}}$). Using n = number of bolts and m = number of faying surfaces, this expression is

$$P_s = DmnT_{i_{\text{spec}}}k_{s_{\text{mean}}} \tag{5.26}$$

where D is a multiplier that provides the relationship between $k_{s_{\text{mean}}}$ and k_s, incorporates the expected bolt tension value as compared with the specified minimum value, and reflects the slip probability level chosen.

Table 12.9. Reduction Factors D for Evaluation of Permissible Shear Loads for Slip-Resistant Coated Surfaces, Turn-of-Nut Installation

| Surface Treatment | Slip Coefficient | | Slip Probability | | | |
| | Average | Standard Deviation | A325 | | A490 | |
			5%	10%	5%	10%
Hot-dip galvanized	0.18	0.04	0.793	0.924	0.755	0.870
Hot-dip galvanized, treated, wire brushed or blasted	0.40	0.07	0.895	1.000	0.848	0.943
Vinyl treated ($t >$ 2 mils)	0.19	0.02	1.072	1.139	1.015	1.072
Blast-cleaned, zinc-sprayed ($t > 2$ mils)	0.40	0.04	1.038	1.110	0.985	1.048
Blast-cleaned, Al-sprayed ($t > 2$ mils)	0.55	0.06	1.040	1.111	0.985	1.047
Blast-cleaned, organic zinc-rich paint	0.35	0.04	1.040	1.111	0.983	1.046
Blast-cleaned, inorganic zinc-rich paint	0.50	0.05	1.040	1.110	0.984	1.048

Of course, the slip resistance of slip-resistant joints with coated faying surfaces can also be expressed using Eq. 5.26. Values of D for use in this equation are tabulated in Table 12.9 for various surface treatments, bolt types, and slip probability levels. The values given in Table 12.9 are for installations in which the turn-of-nut method is used. Table 12.10 gives the values of D to be used in Eq. 5.26 when the calibrated wrench method of installation is used.

The alternative way given in Chapter 5 for expressing the slip load was to derive an equivalent (ficticious) shear stress (Eq. 5.30). This shear stress can then be used in the expression

$$P_s = mn\tau_a A_b$$

where A_b is the cross-sectional area of the bolt corresponding to its nominal diameter, and τ_a can be obtained from Table 12.11 (turn-of-nut installation) or Table 12.12 (calibrated wrench installation), as appropriate.

Chapter 5 should also be consulted about the reductions required if slotted holes

Table 12.10. Reduction Factors D for Evaluation of Permissible Shear Loads for Slip-Resistant Coated Surfaces, Calibrated Wrench Installation

Surface Treatment	Slip Coefficient		Slip Probability A325/A490	
	Average	Standard Deviation	5%	10%
Hot-dip galva-nized	0.18	0.04	0.696	0.792
Hot-diped galva-nized, treated, wire brushed or blasted	0.40	0.07	0.784	0.862
Vinyl treated (t > 2 mils)	0.19	0.02	0.943	0.987
Blast-cleaned, zinc-sprayed (t > 2 mils)	0.40	0.04	0.914	0.963
Blast-cleaned, Al-sprayed (t > 2 mils)	0.55	0.06	0.915	0.964
Blast-cleaned, or-ganic zinc-rich paint	0.35	0.04	0.914	0.962
Blast-cleaned, in-organic zinc-rich paint	0.50	0.05	0.914	0.963

Table 12.11. Permissible Shear Stresses for Slip-Resistant Coated Surfaces, Turn-of-Nut Installation

Surface Treatment	Slip Coefficient		Slip Probability			
			A325		A490	
	Average	Standard Deviation	5%	10%	5%	10%
Hot-dip galvanized	0.18	0.04	9.3	10.8	10.9	12.6
Hot-dip galvanized, treated, wire brushed or blasted	0.40	0.07	22.9	25.5	26.7	29.7
Vinyl treated ($t >$ 2 mils)	0.19	0.02	13.3	14.1	15.5	16.4
Blasted-cleaned zinc-sprayed ($t > 2$ mils)	0.40	0.04	26.5	28.3	31.0	33.0
Blast-cleaned, Al-sprayed ($t > 2$ mils)	0.55	0.06	36.5	39.0	42.6	45.4
Blast-cleaned, organic zinc-rich paint	0.35	0.04	23.2	24.8	27.1	28.8
Blast-cleaned, inorganic zinc-rich paint	0.50	0.05	33.2	35.4	38.7	41.2

are used and for the requirements of a slip-resistant connection under load factor design.

When slip-resistant joints are subjected to sustained loading conditions, only surface treatments that provide adequate slip resistance under long-term loading should be used. Metallizing with either zinc or aluminum or a zinc silicate paint or vinyl coating should be used. Hot-dip galvanizing and organic zinc-rich paint systems are not satisfactory for slip-resistant joints. Obviously, ribbed bearing bolts would be satisfactory for these conditions since they would not permit substantial slips to develop.

If a joint is subjected to repeated loads, the design recommendations given in Section 5.4 are applicable. If the slip resistance is adequate to prevent slip during the lifetime of the structure, the stress range on the gross section area may be used for design. If slip is expected, the design stress range should be applied to the net section. Although several hot-dip galvanized joints have exhibited gross section failures in tests, it is recommended that these connections be designed on the basis of their net section area.

Table 12.12. Permissible Shear Stresses for Slip-Resistant Coated Surfaces, Calibrated Wrench Installation

Surface Treatment	Slip Coefficient		Slip Probability			
			A325		A490	
	Average	Standard Deviation	5%	10%	5%	10%
Hot-dip galvanized	0.18	0.04	8.2	9.3	10.1	11.5
Hot-dip galvanized, treated, wire brushed or blasted	0.40	0.07	20.0	22.0	24.7	27.1
Vinyl treated ($t >$ 2 mils)	0.19	0.02	11.7	12.2	14.4	15.1
Blast-cleaned, zinc-sprayed ($t > 2$ mils)	0.40	0.04	23.3	24.6	28.8	30.3
Blast-cleaned, Al-sprayed ($t > 2$ mils)	0.55	0.06	32.1	33.8	39.6	41.7
Blast-cleaned, organic zinc-rich paint	0.35	0.04	20.4	21.5	25.2	26.5
Blast-cleaned, inorganic zinc-rich paint	0.50	0.05	29.2	30.7	36.0	37.9

Since the presence of a coating does not affect the ultimate strength of a joint, the design recommendations given in Chapter 5 for joints that are not slip critical can be applied to all types of coated joints as well.

REFERENCES

12.1 Office of Research and Experiments of the International Union of Railways (ORE), *Coefficients of Friction of Faying Surfaces Subjected to Various Corrosion Protective Treatments*, Report 2, ORE, Utrecht, the Netherlands, June 1967.

12.2 ORE, *Effects of Weathering on the Coefficients of Friction of Unprotected and Protected Faying Surfaces*, Report 3, ORE, Utrecht, the Netherlands, October 1968.

12.3 ORE, *Influence of Sustained Loading on the Slip Behavior of High Strength Bolted Joints*, Report 4, ORE, Utrecht, the Netherlands, October 1969.

12.4 ORE, *Influence of Coated Surfaces on the Fatigue Strength of High Strength Bolted Joints*, Report 5, ORE, Utrecht, the Netherlands, October 1970.

12.5 Centre de Recherches Scientifiques et Techniques de L'Industrie des Fabrications

Metallique (CRIF), Section Construction Metallique, *Les Assemblages par Boulons de Haute Resistance*, Report MT 48, MT50, Brussels, Belgium, 1969 (in French).

12.6 T. v. d. Schaaf, *Influence of Protective Surface Coatings on the Slip Behavior of High Strength Bolted Joints Subjected to Sustained Loading*, Stevin Laboratory, Report 6-68-5-VB-18, Delft University of Technology, Delft, the Netherlands, 1968 (in Dutch).

12.7 S. Y. Beano and D. D. Vasarhelyi, *The Effect of Various Treatments of the Faying Surface on the Coefficient of Friction in Bolted Joints*, University of Washington, Department of Civil Engineering, Seattle, December 1958.

12.8 L. Zennaro, *Slip Tests of High Strength Bolted Joints with Different Galvanized Coating Structures*, Document CECM-X-71-8.

12.9 V. Lobb and F. Stoller, "Bolted Joints Under Sustained Loading," *Journal of the Structural Division, ASCE*, vol. 97, ST3, March 1971.

12.10 D. J. L. Kennedy, *High Strength Bolted Galvanized Joints*, Engineering Extension Series 15, Proceedings, ASCE Specialty Conference on Steel Structures, University of Missouri, Columbia, 1970.

12.11 American Welding Society, *Recommended Practices for Metallizing with Aluminum and Zinc for Protection of Iron and Steel*, American Welding Society C2.2-67, New York, 1967.

12.12 Steel Structures Painting Council (SSPC), *Guide to Zinc-Rich Coating Systems*, SSPC Specification 12.00.

12.13 W. H. Munse and P. C. Birkemoe, *High Strength Bolting of Galvanized Connections*, The Australian Institute of Steel Construction and the Australian Zinc Development Association, Sydney-Melbourne, Australia, August 1969.

12.14 H. Fouad, "Slip Behavior of Bolted Friction-Type Joints with Coated Contact Surfaces," M.S. Thesis, The University of Texas at Austin, January 1978.

12.15 M. G. Nanninga, "Creep Behavior of Friction-Type Joints with Coated Faying Surfaces," M.S. Thesis, The University of Texas at Austin, January 1978.

Chapter Thirteen

Eccentrically Loaded Joints

13.1 INTRODUCTION

In eccentrically loaded joints, the connection is subjected to applied loads that result in a line of action passing outside the center of rotation of the fastener group. Some common examples are bracket-type connections, web splices in beams and girders, and the standard beam connections shown in Fig. 13.1. Because of the eccentricity of the applied load, the fastener group is subjected to a shear force and a twisting moment. Both the moment and the shear force result in shear stresses in the fasteners, and both of these effects have to be considered in determining the capacity of the connection.

The effect of an eccentric load on a fastener group was studied as early as 1870.[13.1] For a very long period following that, design was carried out on an elastic basis, assuming that rotation of the connection took place about the center of gravity of the fastener group. This meant that the problem could be treated as the superposition of a concentric shear case and a shear due to the torsional moment. The assumption of rotation about the center of gravity was the basis for design tables in the AISC *Manual of Steel Construction* published in 1970,[13.2] although it was recognized that the method produced conservative results, and empirically derived adjustments were permitted.

The assumption of connection rotation about the center of gravity identifies fastener forces that are not compatible with the deformations necessary for such a rotation. An article published in 1914[13.3] reported on work by P. Gullander of Chalmers Technical University at Gothenburg, Sweden that suggested that rotation had to be considered as occurring about an instantaneous center. Calculations done on this basis produced a set of forces in the fasteners that were consistent with the deformations imposed on those fasteners.

The first application of the instantaneous center concept to test results, including use of a measured load versus deformation response for the fasteners, appears to have been made by Yarimci and Slutter,[13.4, 13.5] who performed tests on riveted connections. Since then, both experimental and analytical work has been carried out on bolted connections,[13.6–13.10] with the main effort directed toward information

217

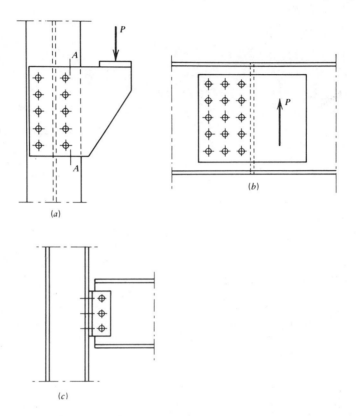

Fig. 13.1. Typical eccentrically loaded connections. (*a*) Bracket connection. (*b*) Beam web splice. (*c*) Standard beam connection.

on the ultimate strength of such connections. In this chapter dealing with the analysis and design of eccentrically loaded fastener groups, emphasis will be placed upon the connection type shown in Fig. 13.1*a*. The application to web splices in girders (Fig. 13.1*b*) and standard beam connections is discussed in Chapters 16 and 18, respectively.

13.2 BEHAVIOR OF A FASTENER GROUP UNDER ECCENTRIC LOADING

Tests on special connections have been performed to evaluate the load versus deformation behavior of fastener groups subjected to an eccentric shear load. Riveted as well as high-strength bolted connections have been examined.[13.4–13.8] Most test specimens were of the type shown in Fig. 13.2, with a fastener group consisting of one or two vertical lines of fasteners. Since the connection is symmetric with respect to the line of action of the load, each test provides two load versus deformation curves for identical connections.

In general, the design of the test specimens caused the fastener group in the

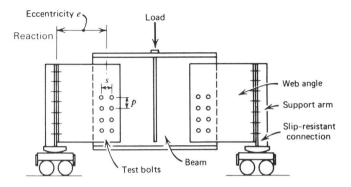

Fig. 13.2.   Test specimen with eccentrically loaded fastener group.

web angles to be the critical component, and the test results can therefore be used to assess the strength of the group. However, the load versus deformation behavior of an eccentrically loaded connection in the field may also be affected by other components of the connection.

The behavior of various fastener patterns under different eccentricities can be represented by load versus rotation curves of the type shown in Fig. 13.3.[13.5] The straight line from the origin to point A represents the elastic rotation, and the transition segment AB identifies elastic as well as plastic deformations. Beyond point B the rotation is mainly produced by plastic deformations. This segment of the load versus rotation curve is terminated by the failure load, reached as one or more of the fasteners fail in shear.

Load versus rotation curves have been developed for bolted as well as riveted specimens.[13.8] Figure 13.4 shows a typical load versus rotation curve taken from the source material of Ref. 13.8 for a bolted specimen with two vertical rows of 3/4-in. dia. A325 bolts. The horizontal distance from the load to the centroid of the fastener group was 12 in. In this test series, the bolt holes in the beam web and web angles were match drilled for fitted bolts. The resulting minimum clearance between the bolt and the hole minimized the joint slip, and caused the applied load

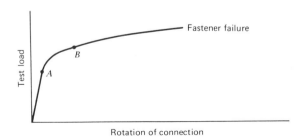

Fig. 13.3.   Idealized load versus rotation diagram for an eccentrically loaded fastener group.

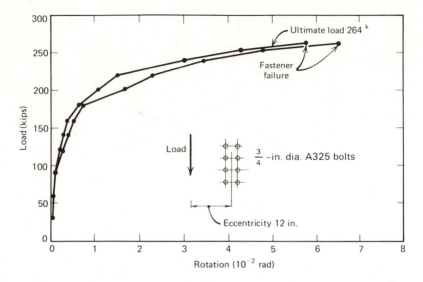

Fig. 13.4. Load versus rotation curve for bolted connections (Ref. 13.8). (Test specimen shown in Fig. 13.2.)

to be taken immediately in bearing on the bolts. In practice, bolts are usually placed in holes with 1/16 in. clearance, and slip may occur when the slip resistance of the connection is exceeded. Slip will bring one or more fasteners into bearing. Thereafter, the connection will behave in much the same way as described by Fig. 13.4.

The amount of slip to be expected depends on the hole clearance, the fastener pattern, and the alignment of the holes in the connection. The rotation due to slip decreases rapidly with an increase in distance from the outermost fastener to the center of rotation of the bolt group. In most practical situations the slips will be so small that they do not have a significant effect on the serviceability of the structure. Therefore, most joints can be designed on the basis of the ultimate strength of the joint.

13.3 ANALYSIS OF ECCENTRICALLY LOADED FASTENER GROUPS

For many years the analysis and design of eccentrically loaded fastener groups was based on the assumption that rotation of the connection takes place about the center of gravity of the fastener group and that the load versus deformation response of an individual fastener is linear.[13.11] In this method, the eccentric load is resolved into a shear load P acting through the centroid of the fastener group and a torsional moment Pe, where e is the eccentricity of P with respect to the centroid of the fastener group. The shear force acting through the centroid is assumed to be distributed uniformly among the fasteners, as in other shear splices. The moment is assumed to cause stresses in the fasteners that vary linearly with the distance from the fastener to the center of gravity. The stress in any fastener is evaluated by vectorially adding the stress resulting from each load component, that is, the concentric shear force and the moment. The method further assumes the connected plates to be rigid enough to remain essentially undeformed during twist. The influence of the frictional resistance between the component parts of the connection is neglected.

Tests on eccentrically loaded riveted connections indicated that the elastic analysis yielded a conservative design.[13.4, 13.5] On the basis of test results, the method was modified by introducing an "effective eccentricity," which is less than the actual eccentricity. Empirical formulas to determine the effective eccentricity as a function of specific fastener patterns were developed.[13.4, 13.5] Reduction in eccentricity yielded a factor of safety more compatible with the value used for shear alone. The method is essentially based on the elastic behavior of the fastener group described in this section. Reducing the eccentricity decreases the magnitude of the bending component and recognizes the actual strength of the joint observed in tests.

Although use of the method just outlined produced safe designs using either the actual or the so-called effective eccentricity, physical testing showed that the factor of safety with respect to the ultimate load was both variable and larger than that used for other types of bolted connections. Furthermore, the method of analysis did not provide the necessary information required in order to accommodate load factor design. As a result, the work described in Section 13.2 was carried out.[13.8] With slight modification, the procedure can be used for the analysis of eccentrically loaded bolted connections in which slip resistance is desired.

13.3.1 Slip-Resistant Joints

Initially, the load versus deformation curve of an eccentrically loaded joint can be approximated by a straight line representing the elastic rotation. (See Figs. 13.3 and 13.4). During this stage the applied load is completely carried by frictional resistance between the constituent parts of the connection. This phase of load transfer is expected to be terminated by a slip of the connection, although physical tests show that slip may or may not occur.[13.12] If slip does occur, the movement that takes place can be expected to bring one or more bolts into bearing. If slip does not occur, it means that one or more bolts were initially in bearing.

A prediction of the load that will cause slip can be made on the basis of the following assumptions:

1. At any value of the load, the connection rotates about an instantaneous center of rotation.
2. At the slip load of the connection, the maximum slip resistance of each individual fastener is reached. An analogous assumption has been used to describe the slip resistance of simple shear splices.
3. The slip resistance of each fastener can be represented by a force at the center of the bolt acting perpendicularly to the radius of rotation.

The instantaneous center of rotation is that point about which pure rotation of the connected parts takes place (See Fig. 13.5.) The term "instantaneous" is used because, in general, the center of rotation is at a different location for each value of the applied load. The location also depends on the fastener arrangement. As shown in Fig. 13.5, the instantaneous center of rotation is located on a line perpendicular to the line of action of the load. This perpendicular must also pass through the center of gravity of the fastener group.

The maximum slip resistance R_s of a single fastener was described in Chapter 5 and can be expressed as

$$R_S = mk_s T_i \tag{13.1}$$

Therefore, based on the previously stated assumptions, at the slip load of the connection each fastener is subjected to a load R_s acting perpendicularly to the radius of rotation. Figure 13.6 shows schematically the load transfer for a symmetric fastener pattern. The three equations of equilibrium must be employed to determine the coordinates of the instantaneous center and the maximum value of the load that results in slip of the connection. The solution of this problem is generally accomplished by an iterative procedure. A trial location of the instantaneous center can be selected. For convenience, the origin of the coordinate system can be placed at the instantaneous center, with the x-axis perpendicular to the applied load. The radius of rotation r_i of the ith fastener is equal to

$$r_i = \sqrt{x_i^2 + y_i^2} \tag{13.2}$$

Equating the sum of all forces in the x and y direction as well as the sum of the moments about the instantaneous center to zero, yields

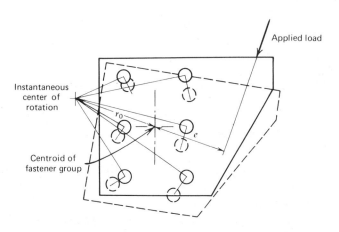

Fig. 13.5.   Instantaneous center of rotation.

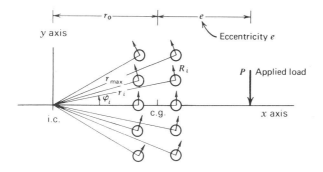

Fig. 13.6. Analyses of eccentrically loaded fastener group. c.g.: Center of gravity of fastener group. i.c.: Instantaneous center of rotation. For slip-resistant joints R_i is equal to R_s where $R_s = m k_s T_i$. For other joints $R_i = (r_i/r_{max}) R_{ult}$.

$$\sum_{i=1}^{n} R_S \sin \varphi_i = 0 \tag{13.3}$$

$$\sum_{i=1}^{n} R_S \cos \varphi_i - P = 0 \tag{13.4}$$

$$P(e + r_0) - \sum_{i=1}^{n} r_i R_S = 0 \tag{13.5}$$

Equations 13.3 and 13.4 are usually written as follows:

$$R_S \sum_{i=1}^{n} \frac{y_i}{r_i} = 0 \tag{13.6}$$

$$R_S \sum_{i=1}^{n} \frac{x_i}{r_i} - P = 0 \tag{13.7}$$

The solution to the problem is achieved if the value of r_0 satisfies all three equilibrium equations. The procedure must be repeated until this condition is met.

A symmetric fastener pattern was used in Fig. 13.6, and the applied load was normal to the axis of symmetry. In such situations the instantaneous center of rotation must lie on the axis perpendicular to the applied load in order to satisfy Eq. 13.7. The procedure also applies to the more general case where no axis of symmetry of the fastener group exists or the applied load is acting in an arbitrary direction, as in Fig. 13.5.

13.3.2 Ultimate Strength Analysis

A theoretical approach to predict the ultimate strength of an eccentrically loaded fastener group was developed by Crawford and Kulak.[13.8] This approach considers the load versus deformation response of a single fastener as the basis for determining the ultimate strength of a fastener group. For a single fastener loaded in double shear, this relationship has been expressed as[5.22]

$$R = R_{ult}(1 - e^{-\mu\Delta})^\lambda \qquad (13.8)$$

in which R = shear force on the bolt at any given deformation, Δ
$\quad\quad R_{ult}$ = the ultimate shear load of the fastener
$\quad\quad\quad \Delta$ = the shearing, bending, and bearing deformation of the fastener as well as the local bearing deformation of the connecting plates
$\quad\mu, \lambda$ = regression coefficients
$\quad\quad\; e$ = base of natural logarithms

Numerical values for R_{ult}, λ, and μ for various combinations of bolts and connected material can be determined experimentally by means of special shear tests. A tension-type shear test has been recommended, since it yields a lower bound to the ultimate shear capacity R_{ult} of the bolt.[4.4]

The evaluation of the ultimate strength of an eccentrically loaded fastener group is comparable to the analysis of similar slip-resistant joints (See Subsection 13.1.1). The connection is assumed to rotate about an instantaneous center, and the connected plates are assumed to remain rigid during this rotation. The latter assumption implies that the deformation occurring at each fastener varies linearly with its distance from the instantaneous center. The fastener deformation and the resulting shear load on the fastener act perpendicularly to the radius of rotation of the fastener. The ultimate strength of the fastener group is assumed to be reached when the ultimate deformation of the fastener farthest away from the instantaneous center is reached.

For a given fastener configuration and an eccentricity of the load equal to e, a trial location of the instantaneous center can be selected at a distance r_0 from the centroid of the fastener group (see Fig. 13.6). The radius of rotation r_i of the ith fastener is given by Eq. 13.2. At the ultimate load of the entire connection, the shear deformation of the critical fastener, located a distance r_{max} away from the instantaneous center, is assumed to be equal to Δ_{max}, the maximum fastener deformation obtained from a single bolt shear test.[4.4,5.22] The deformation of other fasteners can then be determined from

$$\Delta_i = \frac{r_i}{r_{max}} \Delta_{max} \qquad (13.9)$$

The fastener load corresponding to Δ_i is readily obtained from Eq. 13.8.
Equilibrium of horizontal and vertical forces yields

$$\Sigma F_x = 0; \quad \sum_{i=1}^{n} R_i \sin \varphi_i = 0 \qquad (13.10)$$

$$\Sigma F_y = 0; \quad \sum_{i=1}^{n} R_i \cos \varphi_i - P = 0 \qquad (13.11)$$

The summation of moments around the instantaneous center yields a third equation

$$P(e + r_0) - \sum_{i=1}^{n} r_i R_i = 0 \qquad (13.12)$$

Equations 13.10 and 13.11 can be conveniently written in terms of the coordinates x_i, y_i of the fastener,

$$\sum_{i=1}^{n} \frac{R_i y_i}{r_i} = 0 \qquad (13.13)$$

$$\sum_{i=1}^{n} \frac{R_i x_i}{r_i} - P = 0 \qquad (13.14)$$

The solution is obtained when the trial value of r_0 satisfies Eqs. 13.12, 13.13, and 13.14 simultaneously.

13.4 COMPARISON OF ANALYTICAL AND EXPERIMENTAL RESULTS

The validity of the ultimate strength analysis has been checked by comparing predicted results with experimental data. It was found that the predicted ultimate loads for bolted specimens ranged between 5 and 14% higher than the observed failure loads of the connections.[13.8]

One of the reasons for this observed difference is that the deformation of the critical fastener in the connection does not reach the maximum value observed in a single bolt shear test. In the single bolt test, the load and deformation direction do not change throughout the test. In the eccentrically loaded connection, the load and deformation of each bolt are changing direction continuously as the instantaneous center moves with an increase in applied load. It was observed from tested specimens that the bolt holes were deformed and scored by the circular movement of the bolts relative to the plates. Hence, it is unlikely that the critical fastener in the connection will deform as much as a single fastener loaded with a unidirectional force.[13.8]

The predictions of the ultimate strength in Ref. 13.8 were based on load versus deformation relationships determined from compression-type specimens. However, failure of the fasteners was observed mainly in the tension region of the plates (where the connected plates are subjected to tension). It was shown in Chapter 4 that a tension-type shear test generally yields lower shear values than a compression-type shear test.[4.4] Since a compression-type shear test was used by Crawford and Kulak, this may also have contributed to the overestimation of the ultimate loads of the bolt groups reported in Ref. 13.8.

Only a few test results are available for eccentrically loaded slip-resistant connections.[13.12] Figure 13.7 summarizes the load versus rotation curves for three eccentrically loaded connections fastened by $\frac{3}{4}$-in. dia. A325 bolts. The bolts were installed by the turn-of-nut method in holes $\frac{1}{16}$-in. greater in diameter than the bolts. The faying surfaces were in a clean mill scale condition.

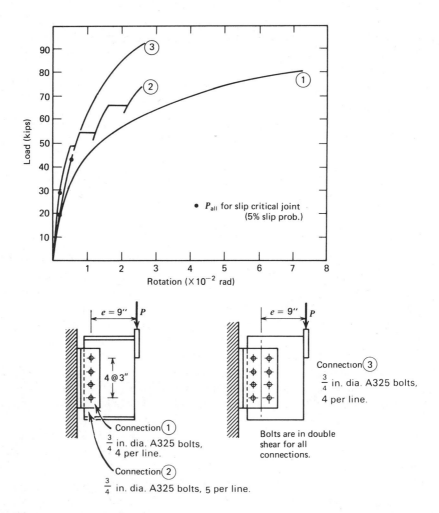

Fig. 13.7. Comparison of design recommendations and test data for slip-resistant eccentrically loaded joints.

Predicted slip loads for these specimens were 30, 44, and 63 kips for connections 1, 2, and 3, respectively. These predictions were based on measured properties of the connection, including both the slip coefficient and the clamping force. Neither connection 1 nor connection 3 showed any slip as the specimens were loaded. Connection 1 did show an increased rate of rotation starting at a load level of about 50 kips. Connection 3 had a fairly stiff load versus rotation response throughout. Connection 2 had an initial slip at a load level of 48 kips and additional slips at 53 and 63 kips. The solid circular dots in Fig. 13.7 indicate the permissible slip load for each specimen using the method outlined in this chapter. A slip probability level of 5% was selected (see Section 5.4).

Although the ultimate strength of an eccentrically loaded fastener group of a type as given in Figs. 13.1a and 13.2 can be evaluated within acceptable limits, additional research is needed to be able to predict the load versus deformation behavior of such joints. Furthermore, research on other types of connections, such as shown in Fig. 13.1b, is desirable to verify the application of the analysis as outlined in the previous section to these types of connections as well.

13.5 DESIGN RECOMMENDATIONS

13.5.1 Connected Material

The design of the plates used in eccentrically loaded joints does not involve special design recommendations. To design the plate for the bracket connection shown in Fig. 13.1a, the shear stress and normal stress at section AA due to the applied load P should be checked. If relatively thin plates are used, the out-of-plane deformations due to instability effects may require an increased plate thickness.

The allowable stresses for these conditions depend on the plate material and the type of loading.

13.5.2 Fasteners

Depending on the required performance of the joint, the permissible load can be based on either the slip resistance or the ultimate strength of the connection. If the latter basis is used (bearing-type connections), the design must be further distinguished as to either allowable stress design or load factor design.

In all these cases, the method of analysis employing the determination of the instantaneous center of rotation can be used. If the joint is to be slip-resistant, the resistance of each bolt will be taken as that established for the type of fastener, surface condition of the connected material, and slip probability level (see Subsection 5.4.2iii). In a bearing-type connection and under load factor design, the resistance of the most highly loaded fastener should be established at its ultimate value (Subsection 5.4.2ii); if allowable stress design is used, the resistance of the most highly loaded fastener should be that corresponding to the allowable stress for the bolt (Subsection 5.4.2i).

In equation form, the results of the instantaneous center method of analysis can be expressed as

$$P = Cr_v \tag{13.15}$$

in which r_v is the permissible load on a single fastener according to the description above (that is, slip-resistant connection, or bearing-type connection under either allowable stress design or load factor design), and P is the corresponding permissible load on the connection acting at a given eccentricity and for a given fastener arrangement. The nondimensional coefficient C provides the necessary relationship between P and r_v.

In developing the coefficients C for bearing-type connections for A325 and A490 bolts for slip-resistant connections, it was observed that the coefficients did not vary greatly for the various cases. Accordingly, it has become customary to tabulate only one set of coefficients and apply it to all cases of eccentrically loaded connections.[13.13, 13.14] Table 13.1 shows a portion of the tabulated C values given in Reference 13.13. The value of C obtained from Table 13.1 for a given geometry of fasteners and eccentricity of load is to be multiplied by the appropriate individual bolt resistance (i.e., A325 or A490 bolt, slip-resistant or bearing-type joint, allowable stress design or load factor design). The coefficients tabulated were generated taking the double shear strength of an A325 bolt as 74 kips and its ultimate deformation as 0.34 in. These were the values established in Reference 13.8, and they were obtained from tests on bolts whose tensile strength was less than 1% greater than the specified minimum value.

There are a number of ways of calculating the permissible eccentric loads for bolt groups for which the C values are not tabulated. The most obvious way is to work from first principles, using the method outlined in Section 13.3. Polynominal functions that approximate the exact solution are also available for one- and two-fastener lines of bolts.[13.8] Some of these were contained in the first edition of this Guide.[13.15] More recently, procedures have been published that identify a reasonable first choice for the trial location of the instantaneous center and provide other computational shortcuts that decrease the number of iterations required for a solution.[13.9] This is particularly helpful if the load vector is not orthogonal with respect to the centroidal axes of the bolt group.

The C values for rivets and A307 bolts can be developed on the basis of typical load versus deformation curves for those fasteners. For convenience, the permissible loads for connections employing these fastener types can be conservatively estimated using the C value for high-strength bolts.

The margins of safety implied by the use of the instantaneous center method of analysis for eccentrically loaded connections are consistent with those of other types of connections. They will be at about the same level as that for concentrically loaded joints less than 50-in. long; that is, a factor of safety of about 2.0 in allowable stress design and a safety index of about 4.5 in load factor design.

Table 13.1. Coefficients C

Required minimum $C = \dfrac{P}{r_v}$

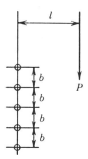

$P = C \times r_v$

n = Total number of fasteners in the vertical row
P = Permissible load acting with lever arm l, inches
r_v = Permissible load on one fastener by Specification
C = Coefficients tabulated below.

							n					
	l (Inches)	2	3	4	5	6 ⫶ 7		8	9	10	11	12
b = 3 in.	3	0.88	1.75	2.81	3.90	4.98	6.06	7.12	8.17	9.20	10.2	11.3
	4	0.69	1.40	2.36	3.40	4.47	5.56	6.64	7.72	8.78	9.84	10.9
	5	0.56	1.15	2.01	2.95	3.98	5.05	6.13	7.22	8.30	9.38	10.4
	6	0.48	0.97	1.73	2.58	3.55	4.57	5.63	6.70	7.79	8.87	9.96
	7	0.41	0.83	1.51	2.28	3.17	4.13	5.15	6.20	7.27	8.36	9.44
	8	0.36	0.73	1.34	2.04	2.85	3.75	4.72	5.73	6.78	7.85	8.93
	9	0.32	0.65	1.21	1.83	2.59	3.42	4.34	5.31	6.32	7.36	8.42
	10	0.29	0.59	1.09	1.66	2.36	3.14	4.00	4.92	5.89	6.90	7.94
	12	0.24	0.49	0.92	1.40	2.00	2.68	3.44	4.27	5.15	6.09	7.06
	16	0.18	0.37	0.70	1.06	1.53	2.06	2.67	3.33	4.06	4.85	5.68
	20	0.15	0.29	0.56	0.85	1.24	1.67	2.16	2.72	3.33	3.99	4.70
	24	0.12	0.25	0.47	0.71	1.03	1.40	1.82	2.29	2.81	3.37	3.99
	30	0.10	0.20	0.37	0.57	0.83	1.12	1.46	1.84	2.27	2.73	3.24
	36	0.08	0.16	0.31	0.48	0.69	0.94	1.22	1.54	1.90	2.29	2.72
b = 6 in.	3	1.39	2.48	3.56	4.60	5.63	6.65	7.65	8.66	9.65	10.7	11.6
	4	1.18	2.22	3.32	4.39	5.45	6.48	7.51	8.52	9.53	10.5	11.5
	5	1.01	1.98	3.07	4.15	5.23	6.28	7.33	8.36	9.38	10.4	11.4
	6	0.88	1.75	2.81	3.90	4.98	6.06	7.12	8.17	9.20	10.2	11.3
	7	0.77	1.56	2.58	3.64	4.73	5.81	6.89	7.95	9.00	10.0	11.1
	8	0.69	1.40	2.36	3.40	4.47	5.56	6.64	7.72	8.78	9.84	10.9
	9	0.62	1.26	2.17	3.17	4.22	5.30	6.39	7.47	8.55	9.61	10.7
	10	0.56	1.15	2.01	2.95	3.98	5.05	6.13	7.22	8.30	9.38	10.4
	12	0.48	0.97	1.73	2.58	3.55	4.57	5.63	6.70	7.79	8.87	9.96
	16	0.36	0.73	1.34	2.04	2.85	3.75	4.72	5.73	6.78	7.85	8.93
	20	0.29	0.59	1.09	1.66	2.36	3.14	4.00	4.92	5.89	6.90	7.94
	24	0.24	0.49	0.92	1.40	2.00	2.68	3.44	4.27	5.15	6.09	7.86
	30	0.20	0.39	0.74	1.13	1.63	2.19	2.83	3.53	4.30	5.12	5.98
	36	0.16	0.33	0.62	0.95	1.37	1.84	2.39	3.00	3.66	4.38	5.18

Source: Manual of Steel Construction, Eighth Edition, American Institute of Steel Construction, Chicago, 1980.

DESIGN RECOMMENDATIONS FOR ECCENTRICALLY LOADED JOINTS

$$P = Cr_v$$

where C = coefficient from Table 13.1, or similar, or as calculated using instantaneous center method of analysis

r_v = permissible load per fastener (kips) accordings to method of design (see the following)

i. Slip-Resistant Connections. If the connection is to be slip-resistant, preloaded high-strength bolts must be used. In this case, the value r_v shall be taken in accordance with Subsection 5.4.2iii, except that the number of fasteners, n, is unity. Thus, from Eq. 5.26,

$$r_v = DmT_{i_{spec}}k_{s_{mean}} \tag{13.16}$$

The resistance per bolt can also be established using the alternative formulation given in Subsection 5.4.2iii.

ii. Bearing-Type Connections: Allowable Stress Design. In the case of bearing-type connections and using allowable stress design, the recommended permissible fastener stresses are 30 ksi for A325 bolts and 40 ksi for A490 bolts (Subsection 5.4.2i). These stresses multiplied by the shear area per bolt (single shear or double shear, as appropriate, will give the resistance value to be used for r_v. If shear planes pass through the bolt threads, a reduction of 70% of the basic value must be used.

iii. Bearing-Type Connections: Load Factor Design. The value of r_v to be used when a bearing-type connection is designed using load factor procedures is given in Subsection 5.4.2ii. For one bolt,

$$r_v = m\phi F A_b \tag{13.17}$$

where m = number of shear planes
ϕ = a reduction factor, 0.80
$F = 0.60\,\sigma_u$ (σ_u = tensile strength of bolt)
A_b = cross-sectional area of bolt corresponding to the nominal diameter

If shear planes pass through the bolt threads, 70% of the value calculated according to Eq. 13.17 should be used.

REFERENCES

13.1 C. Reilly, "Studies of Iron Girder Bridges," *Proceedings of the Institute of Civil Engineers*, Vol. 29, 1870.

13.2 T. R. Higgins, "Treatment of Eccentrically Loaded Connections in the AISC Manual," *Engineering Journal, AISC*, Vol. 8, No. 2, April 1971.

13.3 "Eccentric Rivet Connections," *Engineering Record*, Vol. 70, No. 19, November 7, 1914, p. 518.

13.4 E. Yarimci and R. G. Slutter, *Results of Tests on Riveted Connections*, Fritz Engineering Laboratory, Report 200.63.401.1, Bethlehem, Pennsylvania, April 1963.

13.5 T. R. Higgins, "New Formula for Fasteners Loaded Off Center," *Engineering-News Record*, May 21, 1964.

13.6 A. L. Abolitz, "Plastic Design of Eccentrically Loaded Fasteners," *Engineering Journal, AISC*, Vol. 3, No. 3, July 1966.

13.7 C. L. Shermer, "Plastic Behavior of Eccentrically-Loaded Connections," *Engineering Journal, AISC*, Vol. 8, No. 2, April 1971. (See also discussion by G. L. Kulak, Vol. 8, No. 4, October 1971.)

13.8 S. F. Crawford and G. L. Kulak, "Eccentrically Loaded Bolted Connections," *Journal of the Structural Division, ASCE*, Vol. 97, ST3, March 1971.

13.9 G. D. Brandt, "Rapid Determination of Ultimate Strength of Eccentrically Loaded Bolt Groups," *Engineering Journal, AISC*, Second Quarter, 1982.

13.10 D. M. F. Orr, "The Strength of Eccentrically Loaded Shear Connections," *Journal of Constructional Steel Research*, Vol. 2, No. 1, 1982.

13.11 C. G. Salmon and J. E. Johnson, *Steel Structures, Design and Behavior*, 2nd ed., Harper and Row, New York, 1980.

13.12 G. L. Kulak, "Eccentrically Loaded Slip-Resistant Connections," *Engineering Journal, AISC*, Vol. 12, No. 2, April 1975.

13.13 American Institute of Steel Construction, *Manual of Steel Construction*, 8th ed., AISC, Chicago, Illinois, 1980.

13.14 Canadian Institute of Steel Construction, *Handbook of Steel Construction*, 4th ed., CISC, Rexdale, Ontario, 1985.

13.15 J. W. Fisher and J. H. A. Struik, *Guide to Design Criteria for Bolted and Riveted Joints*, Wiley Interscience, New York, 1974.

Chapter Fourteen
Combination Joints

14.1 INTRODUCTION

Most connections use a single fastening system to connect plates or members together and provide the means of transferring the forces acting in or on the joint. However, situations do arise where it is desirable or necessary to combine two different methods of fastening in a single connection. This generally involves rivets and bolts or bolts and welds. In these connections the two fastening systems share the load. Joints of this type are generally referred to as combination joints or load-sharing joints.

There are two general types of combination connections, as illustrated in Fig. 14.1. One type, shown in Fig. 14.1a, utilizes two different fastening systems to share the load on a common shear plane. This condition may occur when reinforcing or strengthening an existing joint. For example, high-strength bolts may be used to replace several rivets. In other situations, space may not be available for additional fasteners, and welds are added to the joint. In either case, the applied loads are transferred by both types of fasteners on a common shear plane.

Combination joints that combine fasteners on a common shear plane have the advantage of being compact. This reduces the required space and the amount of splice material. In addition, they can help overcome field erection problems. Although welded connections are generally more compact than bolted connections, fabrication tolerances for welding are more rigid than the tolerances allowed for bolted connections. Before the welding process is started, positioning and holding of components in place must also be considered and accounted for. Bolted connections with regular hole clearance ($\frac{1}{16}$-in.) provide for some relative movement between the connected parts after initial assembly and before final tightening of the bolts. Therefore, a member in a frame can be more easily installed with bolts. After the member has been positioned and aligned properly, the bolts are tightened. It is easy to add welds to a connection after it has been first bolted into place (see Fig. 14.1a).

Combination joints of the type as shown in Fig. 14.1a have a wide application for reinforcement of existing mechanically fastened joints. Simple shear splices or

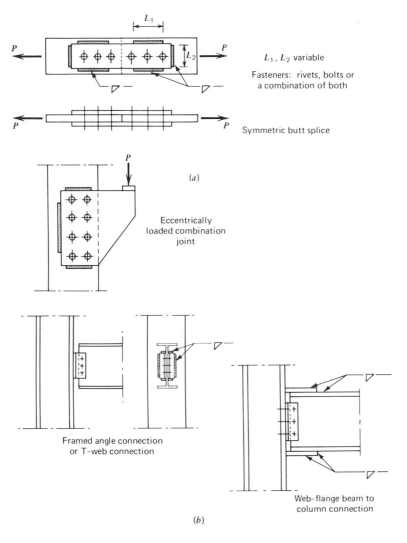

Fig. 14.1. Typical combination joints. (*a*) Load sharing on a common shear plane. (*b*) Combination joints with two different shear planes.

eccentrically loaded shear splices are typical connections that can utilize a combination of mechanical fasteners and welds on a common shear plane.

The behavior of small combination joints with bolts and welds or with bolts and rivets combined on a single shear plane has been studied to evaluate joint behavior and develop design recommendations.[5.5, 9.2, 14.1, 14.2] These tests have demonstrated the applicability of this type of joint. The work in this area is not extensive, and further research would be desirable.

In the other major type of combination connection, two different fastening methods are used but they do not act on a common shear plane. Examples of this category of combination joint are shown in Fig. 14.1*b*. These connections include the simple combination framed beam connection that utilizes shop welds to connect the web angles to either the beam web or the member into which the beam frames and bolts for the field connection. In this particular case, both the bolts and the welds are resisting the beam shear force. Variations of this type of combination joint are possible, such as welding the flanges of beam to column joints and providing a bolted shear connection for the web.

Usually, this type of combination joint will provide greater economy and increased flexibility during erection as compared with the same joint configuration that uses only one type of fastener. The many possibilities for combination joints that exist will only be limited by the ingenuity of the engineer. All available evidence shows that they provide a satisfactory joint with adequate strength and stiffness when proper design procedures are used for the component parts.[14.4]

The remainder of this chapter discusses the behavior of bolted-welded and riveted-bolted combination joints where the fasteners are sharing the load on common shear plane. Other combinations of fastening systems are not considered for this type of combination joint because of the lack of information and because of their limited use in structural applications.

Discussion of the behavior of the other major type of combination connection, where different types of fasteners are used but not on a common shear plane, is given in Chapter 18.

14.2 BEHAVIOR OF COMBINATION JOINTS THAT SHARE LOAD ON A COMMON SHEAR PLANE

Before the combined action of two different fastening methods acting in a common shear plane is discussed, it is desirable to reexamine the load versus deformation behavior of the different types of individual fasteners. Figure 14.2 shows typical load versus deformation curves for welded, bolted, and riveted tension specimens. This figure indicates that high-strength bolted connections with normal hole clearance provide a very high initial stiffness up to the slip load of the connection. During slip, the deformations increase significantly until the bolts come into bearing. After the bolts are in bearing, the load versus deformation curve shows an increase in joint stiffness. Joint slip can be minimized by installing fitted bolts in matching drilled holes.

Compared to slip-resistant high-strength bolted joints where the load is transferred by friction, riveted connections are generally more flexible. Often a sudden change in the slope of the load versus deflection curve can be observed that is directly comparable to slip in a high-strength bolted connection. However, this "slip" is usually less than one-third the slip observed in high-strength bolted connections.

A typical characteristic of a welded connection as compared with riveted or

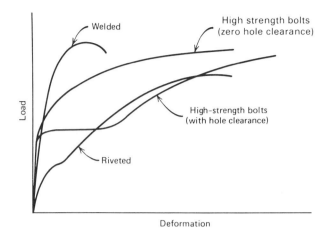

Fig. 14.2.   Load versus deformation relationships for different fastening methods (Ref. 9.2.).

high-strength bolted connections is the reduced deformation capacity. Slip does not occur in welded connections, and the initial stiffness of the joint only changes as the ultimate load is approached. From these load versus deformation relationships for typical fasteners, one can conclude that combination of these fasteners would be most appropriate where compatible deformation characteristics exist. The preferred combinations appear to be welds with slip-resistant high-strength bolts and rivets with bolts.

14.2.1   High-Strength Bolts Combined with Welds

A comparison of the load versus deformation capacity of welded and high-strength bolted connections with normal $\frac{1}{16}$-in.-hole clearance indicates that the total deformation capacity of the welds is of the same order of magnitude as the maximum slip of a high-strength bolted connection. Therefore, if both fastening methods are used on a common shear plane, the capacity of the resulting combination joint might be taken as the sum of the weld strength and the slip resistance provided by the bolts.

The question arises as to what constitutes failure in a welded-bolted combination joint. As discussed above, because the weld shear deformation capacity and the observed values of slip in bolted joints are about the same, the weld shear failure can be expected to occur at the same time as the bolts slip into bearing. If the joint was designed to be slip-resistant, this would constitute failure. If the joint were designed as a bearing type, the connection now consists of a bolted connection (with some broken welds) whose capacity can be determined according to the usual rules (see Chapter 5). Thus, in new work it is not logical to consider using both high-strength bolts and welds in the same joint unless it is categorized as a slip-resistant connection. It must also be noted that in load factor design a slip-resistant connection must also be checked with respect to the ultimate limit state. The ultimate resistance of the bolted-welded joint, as defined by complete separation of the parts, will be the greatest of the shear capacity of the bolts, the bearing capacity of the plates, or the shear capacity of the welds. The resistance so determined must be at least equal to the effect of the factored loads.

In renovation or repair work, two separate loading cases should be identified. If, for example, welds are added to a bolted joint that has little or no load, the case is the same as that described for new work. On the other hand, if the joint is already under load, the existing component, bolts or welds, must be initially carrying that load. Load applied subsequent to the addition of welds or bolts will be shared between the original fastening elements and those that have been added. Whether the joint is to be considered now as slip resistant or bearing type will depend upon individual circumstances. Similarly, if the joint is a bearing type, the identification of the critical fastening element will have to be done on a case-by-case basis.

Tests have been performed to evaluate the validity of the assumption made for new work (or for renovations done under no load), namely, that the shear capacity of the welds and the slip resistance of the bolts can be added.[5.5, 9.2, 14.1, 14.2] The test joints were generally small tension type butt splices with two bolts on either side of the splice, as shown in Fig. 14.3. The influence of the location of the welds, that is, either transverse or parallel to the applied load, was also studied. Furthermore, the ratio of the capacity of the welds with respect to the slip resistance of the bolts was considered as a test variable.

Figure 14.3 summarizes the results observed in a typical series of test joints.[9.2] The load versus deformation behavior of the plain welded and the plain bolted connection is shown, as is that for the combination bolted and welded joint. It is apparent that the behavior of the combination joint can be adequately described by the sum of the slip load of the plain bolted connection and the strength of the welds. Other combinations of weld length, weld location, and slip resistance of the bolted joint resulted in similar conclusions.[9.2, 14.1]

The tests reported in Ref. 9.2 were limited to small connections with only a few bolts in line. In larger connections, some misalignment may exist and the bolts come into bearing before failure of welds occurs. The load carried by the bolted connection is then transmitted by friction and bearing. The failure load of these connections is likely to exceed the estimated ultimate load determined from the slip resistance of the bolts and the strength of the welds. Reducing the hole clearance would also bring the bolts into bearing and increase the ultimate strength of a bolted-welded combination joint. The maximum capacity of a combination joint is developed when fitted bolts are installed in matching drilled holes. Tests have indicated that these connections have an ultimate load that exceeds the summation of the weld strength and the slip load of the bolted connection.[9.2, 14.2] Obviously, such joints are not very economical in new work. However, in existing work holes would of necessity be drilled, and they would of course be matched. In this case, fitted bolts could be easily used.

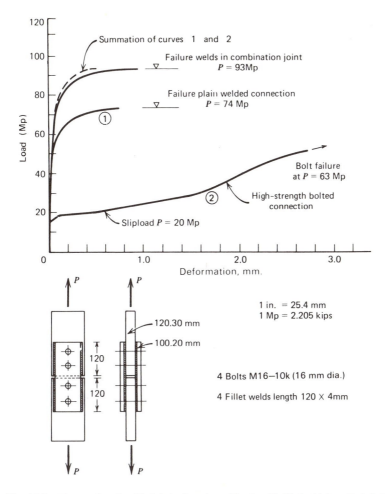

Fig. 14.3. Test results of welded, bolted, and combined welded-bolted joints (Ref. 9.2).

Another aspect that has to be considered is the behavior of combination joints under repeated loading conditions. The behavior of high-strength bolted connections subjected to repeated loading conditions is discussed in Chapter 5. Tests performed in Germany indicated that the fatigue strength of a high-strength bolted connection decreases when weldments are added.[9.2] This reduction in fatigue strength is expected, because the weld toe is the critical region, and crack growth will occur just as in a welded joint. The weld toe was more critical than the bolt holes in all test joints. [9.2] A comparison of the few data available for welded joints indicates that the fatigue strength is not significantly different from the fatigue strength of a similar plain fillet welded connection. Hence, the design criteria for welded joints should be used for cyclic load conditions when the welds are positioned on the boundaries of the combination joint.

Some tests have indicated that an improvement in fatigue strength can result when the welds are placed on the joint interior.[14.3] This removes the weld from the more highly stressed joint boundary where the geometric discontinuity is more severe and places it in a lower stressed region. In addition, the stress concentration condition is generally decreased, since the connected parts are more nearly subjected to about the same strain conditions. However, caution must always be exercised when adding weld to existing bolted joints. The danger exists that conditions favorable to crack growth will be created, particularly if these welds are used at plugs or slots.

14.2.2 High-Strength Bolts Combined with Rivets

A combination of rivets and high-strength bolts intersecting the same shear plane would not be used in new construction. However, high-strength bolts are often used to replace one or more rivets in existing riveted connections. This is done to either repair the joint or to strengthen the connection.

The addition of high-strength bolts to a riveted connection results in a number of improvements. For a given diameter, a high-strength bolt has a greater shear strength than a rivet, and so the ultimate strength of the whole connection will be increased. If the connection is slip-resistant, the stiffness will be increased by the addition of high-strength bolts. If the slip resistance is exceeded, the presence of the rivets, which have less hole clearance than do high-strength bolts, means that less slip will take place as compared with a fully-bolted joint. Furthermore, replacing rivets by high-strength bolts has been shown to improve the fatigue strength of the joint.[14.5]

Tests to evaluate the load versus deformation behavior of short bolted-riveted combination joints have indicated that the ultimate strength of the joint is adequately approximated by the summation of the resistance of the two types of fasteners.[9.2] This is illustrated in Fig. 14.4, where the load versus deformation curves of a riveted, a bolted, and a bolted-riveted combination joint are compared. This figure clearly shows the increased stiffness of the combined joint as compared with the riveted joint. The improved slip behavior of the combination joint is also evident.

Since the joint strength of short combination joints is an aggregate of the strengths of the individual fasteners, it does not matter how the fasteners are arranged in the combination joint. Hence, either the outermost rivets or rivets located in the joint interior can be replaced by high-strength bolts. Either arrangement yields about the same ultimate load. Based upon the observed behavior of long riveted and bolted joints, the fastener location will influence the joint strength. Because of "unbuttoning," replacing the outermost rivets of a long joint by high-strength bolts will be more effective in increasing the joint strength than replacing the same number of interior fasteners. Experimental verification is not available on long joints at the present time (1987).

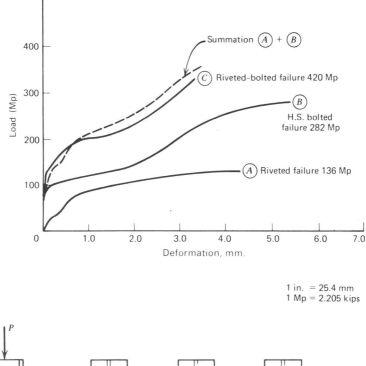

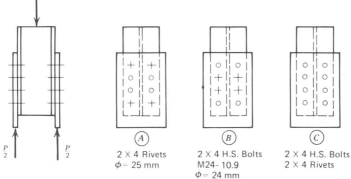

Fig. 14.4. Test results of riveted, bolted, and combined riveted-bolted joints (Ref. 9.2).

As was discussed for the case of bolted-welded joints, the amount of load on a rivet joint at the time that high-strength bolts are added must be known. If the existing load is small or zero, the rivets and bolts can be assumed to share the load, as discussed above. However, if the joint is already under load at the time of reinforcement, the rivets will already be under load when the high-strength bolts are introduced. Additional load will be shared between the rivets and the bolts.

The joint strength will have to be assessed on a case-by-case basis, and this can be done by relating the fastener deformations and loads.

Consider the replacement of some rivets at the extremities of a loaded, riveted joint by high-strength bolts. Under the existing load, the shear force per rivet can be calculated (see Subsection 5.2.5) and the corresponding shearing deformations, Δ_e, established using values similar to those shown in Fig. 3.2. The rivet shearing deformation at ultimate, Δ_m, is also obtainable from Fig. 3.2. When high-strength bolts are added, they can only be subjected to the difference between the two rivet shear deformations, that is, $\Delta_b = \Delta_m - \Delta_e$. The force per bolt can then be obtained from a figure such as Fig. 4.11 or from the mathematical expressions developed to describe this relationship.[5.22] Finally, the ultimate capacity of the riveted-bolted combination joint can be calculated as the sum of the rivet forces and the bolt forces established as above. Although the procedure described is believed to be sound, there have not been any tests that would verify its applicability.

Many test programs have indicated that high-strength bolted shear splices subjected to repeated-type loading generally exhibit a significantly higher number of load cycles before failure than do comparable riveted specimens (see Chapter 5). This difference is mainly attributed to the high clamping force provided by the bolts, which results in a more favorable stress distribution around the bolt hole as compared with the the stress flow around the holes in a riveted connection. Hence, the replacement of rivets by high-strength bolts will increase the fatigue strength of a connection.

Fatigue strength tests have been carried out on both small bolted-riveted combination joints[9.2] and on full-size specimens.[14.5] In the latter program, 16 full-scale tests were conducted, including both modeled joints and actual connections taken from a structure in service. The study showed that the replacement of rivets with preloaded high-strength bolts at locations of observed or anticipated cracking increased fatigue life by a factor of from two to six. Proper removal of the rivets to be replaced and proper installation of the replacement bolts is necessary so that no new mechanical flaws (burrs, nicks, and gouges) are introduced during the rehabilitation process. The tests also showed that if cracking is retarded in the critical region by rivet replacement, other locations not as highly stressed may become critical.

Regression analyses of the data were carried out that enabled the prediction of the fatigue strength of the rehabilitated joints.[14.5] For cases involving structural sizes similar to those tested, these could be used. Alternatively, the conservative prediction might be used; that is, rehabilitation of a joint by replacement of rivets with preloaded high-strength bolts will result in a fatigue life twice as great as that of the unrehabilitated joint.

14.3 DESIGN RECOMMENDATIONS

Although only limited test data are available, a knowledge of the behavior of the different fastener responses enables design recommendations to be developed for combination joints that utilize two different types of fasteners to transfer load on a common shear plane.

14.3.1 Static Loading Conditions

For welded-bolted cases in which the load in the joint to be reinforced is small or zero, the capacity can be taken as the sum of the slip resistance of the high-strength bolted part and the ultimate load of the welded part. This summation corresponds to the slip resistance of the connection. If load factor design is being used, the ultimate resistance (separation of the parts) must also be calculated and compared with the force introduced into the joint by the factored loads. For the welded-bolted joint, this will always be the ultimate shear capacity of the bolts or the bearing capacity of the connected parts.

If the welded-bolted combination joint arises as a result of reinforcement under load, then it must be recognized that the original fastening element is already loaded, and only loads applied after the reinforcing connector is introduced will be shared. The identification of the critical fastening element and the joint resistance will have to be handled on an individual basis, considering the deformation and load responses of the individual elements and enforcing compatibility and equilibrium requirements.

Bolted-riveted combination joints will similarly have to be distinguished as to loading case. If the combination is formed under low or zero load, the rivets and bolts can be assumed to share all the applied load. The capacity of the joint will be the sum of the individual contributions. If reinforcement is made under load, usually by the replacement of rivets with high-strength bolts, then the load and deformation originally present in the rivets must be calculated. The load applied after the reinforcement will be carried by the rivets and bolts in proportion to their deformations.

14.3.2 Repeated Loading Conditions

When high-strength bolts and fillet welds are combined to resist forces on a common shear plane, the fatigue strength is governed by the welded joint. Crack growth occurs first from the weld toe termination, and fatigue provisions for the welded detail should be used for design.

When high-strength bolts have been used to strengthen riveted joints, a significant improvement in fatigue strength has been noted when the bolts were placed at the joint ends where the stressed plates are most critical. Regression analyses are available that will enable the prediction or the fatigue strength of the rehabilitated joints.[14.5] Alternatively, it would be conservative to assume that the rehabilitation of a joint by replacement of rivets with preloaded high-strength bolts will give a fatigue life (at the point of rehabilitation) twice as great as that of the unrehabilitated joint. The possibility that other regions of the connection might now become critical must also be considered.

REFERENCES

14.1 W. Hoyer and H. Skwirblies, *Hochfeste Schrauben in Verbindungen Mit Schweiss-nachten*, (2nd report), Wissenschaftliches Zeitschrift der Hochschule fuer Bauwesen, Cottbus, 1959/1960, Vol. 1, Cottbus, Germany, 1960.

14.2 N. M. Holtz and G. L. Kulak, *High Strength Bolts and Welds in Load-Sharing Systems*, Department of Civil Engineering, Nova Scotia Technical College, Halifax, Nova Scotia, September 1970.

14.3 E. Ypeij, *New Development in Dutch Steel Bridge Buildings*, Preliminary Report 9th Congress IABSE, Amsterdam, May 1972.

14.4 J. S. Huang, W. F. Chen, and J. E. Regec, *Test Program of Steel Beam-to-Column Connections*, Fritz Engineering Laboratory Report 333.15, Lehigh University, Bethlehem, Pennsylvania, July 1971.

14.5 H. S. Reemsnyder, "Fatigue Life Extension of Riveted Connections," *Journal of the Structural Division, ASCE*, Vol. 101, ST12, December 1975.

Chapter Fifteen
Gusset Plates

15.1 INTRODUCTION

When the longitudinal axes of two or more members to be joined at a point are inclined with respect to one another, it is not usually possible to bolt one member directly to another. In these cases, gusset plates are used to receive the load from one member and transfer it to the others. Figure 2.4 illustrates such a connection. Although it is customary to assume that the members in this arrangement are loaded only in their axial directions, the delivery of these loads by the bolts into and out of the gusset plate will produce bending, shear, and normal forces at any arbitrary section taken through the gusset plate.

Out-of-plane bending in gusset plates is generally insignificant. In most cases, the load application is symmetric with respect to the plane of the gusset plate, or joint geometry prevents or minimizes the secondary out-of-plane bending stresses, as shown in Fig. 2.4c. Because of these factors, the analysis of gusset plates generally is treated as a two-dimensional plane stress problem; secondary stresses due to out-of-plane bending are neglected.

Until very recently, there had been relatively few attempts to determine the stress distribution in gusset plates, either analytically or experimentally. The usual procedure [15.1] was to select sections for examination (usually taken parallel to and perpendicular to the chord in the case of a truss), identify the bolt forces that had been delivered to the gusset plate, and use these forces to calculate the shear, normal force, and moment at the cut section. The stresses were then calculated assuming that the elementary formulas for beams apply. It was recognized that the assumption of beam behavior is not valid, however. Furthermore, it was uncertain whether local stresses within the gusset plate necessarily remained elastic, even under allowable stress design. An early study on a model of a Warren truss lower chord connection indicated that the beam assumptions led to erroneous predictions.[15.2] (This study was limited to nominal stresses within the elastic range.)

Since about 1970, more experimental and analytical studies have become available. [15.3–15.9] The latest studies, using the finite element method to model the structure, have been particularly useful in predicting the stresses that occur in the inelastic region of behavior. Although the amount of experimental data is still

small, the analytical studies are able to provide a good prediction of test results.[15.7, 15.9]

This chapter discusses the methods currently in use for the design of gusset
plates. An examination of current practice suggests that substantial variations in
the factor of safety against ultimate load exist in gusset plates because of the
assumptions involved. Despite the shortcomings of the presently available design
methods, these procedures continue to be used because experience with these methods has resulted in gusset plates that have provided satisfactory performance and
behavior. There are, no doubt, substantial variations in the actual strength of the
various gusset plates that result from this design approach. However, there are no
known failures or documented cases of adverse behavior.

15.2 METHOD OF ANALYSIS AND EXPERIMENTAL WORK ON GUSSET PLATES

The design of gusset plates has long been based on simple methods of analysis.
Simple strength of materials analysis or specification rules were used.[15.10, 15.11] Such
an analysis is based on assumptions, and their adequacy is not fully known.

The procedure generally followed and presented in many design handbooks is
summarized as follows.[15.1] It is assumed that all fasteners connecting a member to
the gusset carry an equal share of the load. This permits the number of fasteners
required to transmit the load from each member into the gusset plate to be determined. Note that comparable assumptions regarding the load transfer are used for
design of other types of shear splices. The planar dimensions of the plate are
selected so that all fasteners can be placed. A tentative plate thickness is selected,
often on the basis of experience of the designer or as prescribed by applicable
specifications. Stresses are then evaluated on each section by assuming the plate
to act as a beam. Hence, beam theory is used to evaluate the stesses at the selected
section. Generally, the analysis consists of checking various sections through the
plate in order to obtain the governing one (see Fig. 15.1).

It has long been recognized that the beam method of analysis is of questionable
value.[15.1-15.5] The load partition among fasteners connecting a member to a gusset
plate is generally not uniform, and the applicability of beam formulas to the geometries generally encountered in gusset plates is questionable. To examine the
validity of the use of beam formulas for this problem, Whitmore, in 1952, investigated the stress distribution in a $12.6 \times \frac{1}{8} \times 16.6$ in. aluminum gusset plate in
which the connections were made using tight fitting pins and bolts.[15.2] The model
simulated a lower chord joint of a Warren-type truss with a continuous chord (see
Fig. 15.2). A vertical member was attached to the model but not loaded. Whitmore
observed that the locations of the maximum tensile and compressive stress were
near the ends of the tension and compression diagonals, respectively. The assumption that normal stresses, bending stresses, and shear stresses on a critical
plane through the ends of the diagonals are distributed according to beam formulas

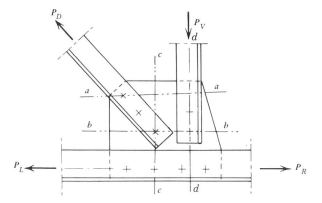

Fig. 15.1. Analysis of gusset plates. Bending stress $\sigma_{max} = P/A \pm Mc/I$. Shear stress $\tau_{max} = 3/2$ V/A. a-a, b-b, c-c, and d-d denote sections to be checked.

was found to be inaccurate. This is illustrated in Fig. 15.3 where the distribution of the vertical normal stress along a section parallel to the chord member and passing through each diagonal is shown. A significant difference between the calculated and observed stresses is noted, particularly at the edges of the plate.

Whitmore concluded that the maximum normal stress at the end of a member

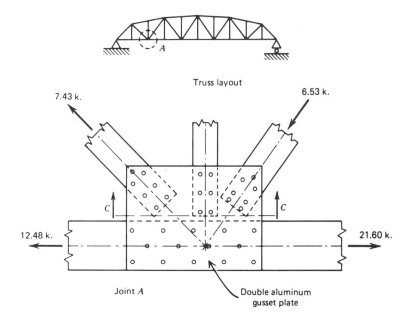

Fig. 15.2. Gusset plate model as used by Whitmore (Ref. 15.2.).

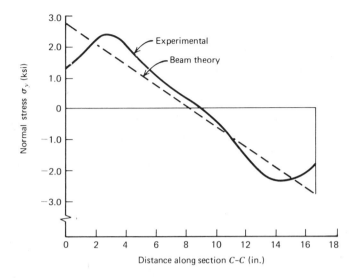

Fig. 15.3. Distribution of vertical normal stress on critical section C-C; see Fig. 15.2.

could be estimated adequately by assuming that the member force was distributed uniformly over an effective area of plate material. This area was obtained by multiplying the thickness of the plate by an effective length. The effective length was estimated by constructing 30° lines from the outer fasteners in the first row to their intersection with a line perpendicular to the line of action of the external load and passing through the bottom row of fasteners, as shown in Fig. 15.4. The segment intercepted by the 30° lines is then used as the effective width of the plate.

Methods of analysis have become available, such as the finite element method, which permit the gusset plate to be analyzed in the elastic and inelastic ranges. Vasarhelyi[15.4] and Davis[15.5] both attempted an elastic finite element solution of

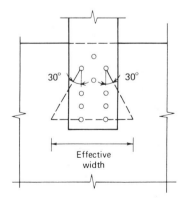

Fig. 15.4. Evaluation of effective width for fastener pattern.

specific gusset plates. Struik[15.6] not only studied the problem in the elastic range but also predicted the behavior of gusset plates in the inelastic range up to their ultimate strength. In the elastic-plastic analysis, the presence of the holes are accounted for in an approximate manner. Richard and his coworkers at the University of Arizona have also modeled gusset plate behavior using the finite element methods.[15.7] Their procedure is particularly interesting because the model includes all elements of the connection: the fasteners, the connected members, and the gusset plate itself.

The elastic analyses [15.4-15.6] confirmed Whitmore's conclusions. Significant variation between stress distributions predicted by the finite element method and beam theory existed. However, the difference was not necessarily unsafe. None of the stresses evaluated by the finite element analyses exceeded the maximum values predicted by beam theory. The location and distribution of the maximum stresses showed substantial variation.

Some of the results of the elastic-plastic finite element analysis done by Struik[15.6] of a typical gusset plate are shown in Figs. 15.5 through 15.7. Figure 15.5 shows the geometry of the gusset plate as well as the applied loads. The tensile strength of the material was assumed to be 70 ksi at a strain of 15%. Reaching the tensile strength in one or more elements was considered to result in failure of the gusset and defined the ultimate load.

The predicted load versus displacement curves for two typical points on the gusset are shown in Fig. 15.6. The elastic-plastic boundaries corresponding to the load levels P_1, P_2, and P_3, indicated in Fig. 15.6, are summarized in Fig. 15.7. It is apparent that yielding occurred near the ends of the members soon after load P_1 was applied. The load versus deformation curves start to deviate from linearity, reflecting plastification of the section. At load stage P_3 the system exhibited sub-

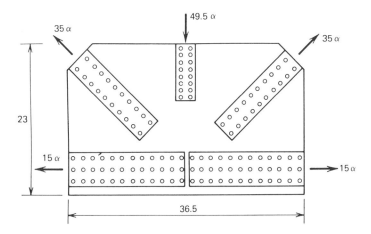

Fig. 15.5. Geometry and loading conditions for sample gusset plate. Fastener holes 0.5-in. dia. Plate thickness 0.25 in. α Load parameter to indicate proportional loading (see Fig. 15.7).

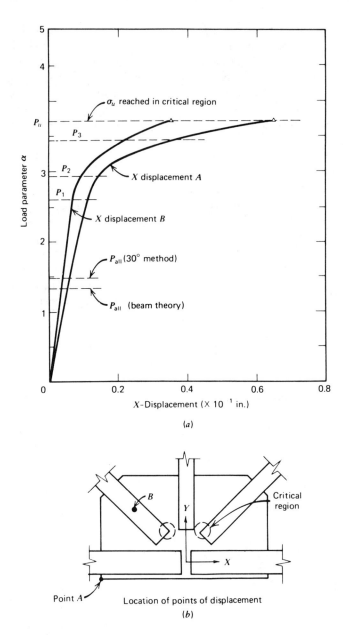

Fig. 15.6. Typical load versus displacement curves for sample gusset plate. (*a*) Load versus displacement curves. (*b*) Location of points of displacement.

248

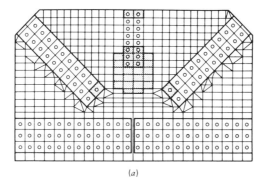

(a)

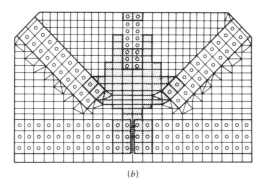

(b)

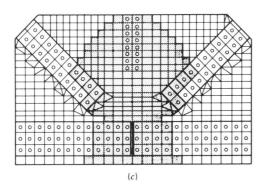

(c)

Fig. 15.7. Elastic-plastic boundary at various load stages (for load reference see Fig. 15.6). (a) Load stage P_1. (b) Load stage P_2. (c) Load stage P_3.

249

stantial nonlinear behavior. The tensile strength was first reached in the elements at the end of the diagonal members, as indicated in Fig. 15.6.

The allowable loads for this particular gusset plate were evaluated on the basis of the current AISC specifications[2.11] and are also shown in Fig. 15.6. The elastic-plastic analysis indicated a factor of safety against ultimate between 2.5 and 2.7, depending on the method of analysis used. For this particular example, the 30° effective width method gave a slightly higher allowable load than beam theory. On the basis of these finite element studies, it was concluded that current design procedures result in a variable factor of safety against the gusset plate capacity.[15.6]

The physical tests done by Bjorhovde at the University of Alberta[15.8] allowed the analytical model of Richard *et al.* to be evaluated.[15.7] The general arrangement of the test specimen is shown in Fig. 15.8. The angle of the bracing used in the tests was 30°, 45° (as shown), or 60°. Gusset plate thicknesses were $\frac{1}{8}$ and $\frac{3}{8}$ in.

A plot of theoretical effective stresses (according to the von Mises yield criterion) is shown in Fig. 15.9 for the case of a $\frac{1}{8}$-in. thick gusset plate loaded by a member inclined at 45° and for a load of 150 kips. This is the load at which the gusset plate in the physical test tore in the region around the ends of the splice plates. The plot shows that a considerable amount of the splice plate had yielded at this load level, and that high strains were present in the region where the failure actually occurred. A comparison of measured strains with those obtained analytically showed reasonable agreement in regions where the strain gradient is low. In areas of high

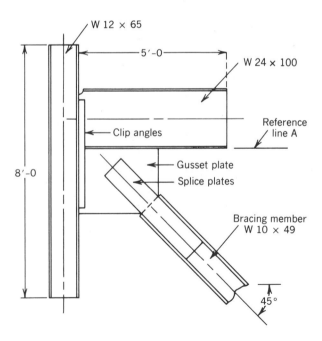

Fig. 15.8.   University of Alberta test specimen.

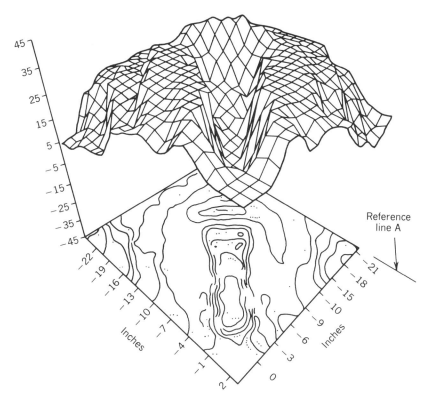

Fig. 15.9. 45° $\frac{1}{8}$-in. gusset, effective (von Mises) and surface plots, 150 kip load.

strain gradient, there were quite large differences between measured and calculated strains, possibly a reflection of inadequate mesh size in the finite element model. Finite element analyses for the 30° and 60° load cases and for the other gusset plate thickness showed similar results.

In the tests done by Bjorhovde,[15.8] the cases in which tearing occurred in the gusset plate in the region of the splice plate ends occurred at loads of 150 kips and 158 kips. Both these cases were for gusset plate $\frac{1}{8}$-in. thick, and the loaded members were inclined at 45° and 30°, respectively. The Whitmore method, described earlier, would give a predicted failure load of 179 kips for each of these tests. (The measured ultimate tensile strength of the gusset plate material was 55.5 ksi.) In working stress design, and using the specified minimum yield strength (44 ksi), the permissible load using the Whitmore method is 142 kips. For these tests, the Whitmore method overestimates the ultimate strength, and it does not provide a sufficient margin of safety if allowable stress design is used.

In another test of a gusset plate in which a 45° loaded member was present, a $\frac{3}{8}$-in. thick gusset plate was used.[15.8] It was loaded to 324 kips before the test had to be discontinued. Although failure had not occurred, a small amount of yielding was observed at the same location where gusset plate tearing took place when the $\frac{1}{8}$-in. thickness was used. Using the measured material yield strength of 42.7 ksi, the Whitmore method would predict yielding at a load of 412 kips.

Richard *et al.*[15.7] have suggested that the block shear method of analysis used for standard web angle connections on coped beams might be suitable for gusset plate design. (This method is discussed in Subsection 18.3.1.) For "long" connections, described as those with more than five bolts in line, they suggest using the gross area along the potential failure surfaces, and for short connections, they suggest using the net area. The connection shown in Fig. 15.8 has nine bolts in line at a spacing of $2\frac{1}{4}$ in. The bolt lines are 5 in. apart. Using the measured ultimate tensile strength (55.5 ksi) and taking the ultimate shear stress as 0.6 times the measured tensile yield strength (42.7 ksi), the calculated ultimate load by this method is 150 kips. This is exactly equal to the failure load of the connection in which the 45° member was present and very close to the failure load of 158 kips for the case when a 60° member was present. More studies are needed to support the block shear model; however, it is consistent with the maximum shear stresses identified in the analytical model of Ref. 15.7, and it seems to give good results.

The design of the gusset plate assembly must also include consideration of the forces in the bolts. In Chapter 5 the evaluation of individual bolt forces along the length of a joint assumed that the two parts being connected (in Chapter 5, lap plates and main plate) would be of equal cross-sectional area. For a connection of the type shown in Fig. 15.8, the "main plate" is the gusset plate and the "lap plates" are the splice plates or main member. It is uncertain as to just what width of gusset plate should be associated with the splice plates at any given location, but it is clear that in these test specimens the amount of splice plate cross-sectional area greatly exceeded that of any associated gusset plate. As such, and using the concepts developed in Chapter 5, it could be expected that nonuniformity of load among fasteners in such an arrangement will be relatively large. Furthermore, this nonequal loading of bolts will not be symmetrical about the midlength of the joint as it was for butt splices. Relatively higher fastener loads can be expected at the end of the connection toward the interior of the assembly.

The relative dimensions used in the connection shown in Fig. 15.8 are probably not typical of fabrication practice; the thin gusset plate was necessary to ensure plate failure. Nevertheless, using the dimensions of the test specimen, Richard *et al.* found that the maximum bolt shear force was about 1.5 times greater than the average bolt shear force for the case of the 60° member. Considering gusset plate thicknesses that are more representative of those used in practice, this inequality would not be quite so large. In any case, the effect of these higher fastener loads is not likely to show up as a problem with respect to a shear failure of the bolts, but will more likely result in high bearing stresses in the gusset plate. There is not enough information at the present time to draw conclusions from this point. Keeping in mind that existing practice has not resulted in any known problems in the behavior of gusset plates, it is probably sufficient to note that special attention should be paid to bearing stresses when long joints (say, more than five bolts in line) are present, and that the rules for minimum end and edge distances for the fasteners should be strictly followed.

As of (1987), no tests or analysis have focused on the behavior of gusset plates when compressive, rather than tensile, loads are delivered by the connected members. Obviously, the presence of one or more compressive loads being delivered to a gusset plate raises the possibility of local buckling in the plate. Good practice indicates that the region of unsupported gusset plate at the ends of the members (see Fig. 15.6b, for example) should be kept to a minimum. If the ends of the members being connected are not in close proximity, the possibility of local buckling must be examined. At the present time, this can only be done on the basis of engineering judgement supplemented by the current state of knowledge of forces delivered to gusset plates by members loaded in tension.

The complexity of a gusset plate connection and the many possible arrangements of boundry members, fasteners, and gusset plate geometries means that much investigative work still needs to be done. The proximity of the member being examined to the other members can undoubtedly affect the ductility of the gusset plate and, thereby, the ultimate load. The designer must be aware of all possible modes of failure in the assembly: local buckling of the gusset, tearing of the gusset, bolt shear failure, and bearing failure in the plate around the bolts. The modern tools of analysis that are now available, in conjunction with additional physical tests, should help to provide more information on these topics and help improve current design procedures.

15.3 DESIGN RECOMMENDATIONS

Design recommendations for gusseted connections concern the fasteners as well as the plate material. To determine the total number of fasteners required to transfer the load from a member into the gusset plate, equal load distribution among the fasteners may be assumed, as is done with other joints. Design recommendations for fasteners are given in Chapter 5 for symmetric butt splices and are applicable to the design of slip-resistant and bearing-type gusset plates as well. Long joints, those in which the number of bolts in a line is greater than five, or joints in which the gusset plate is thin relative to the amount of material being connected, should receive special attention. End and edge distance requirements for the bolts should be strictly followed.

The analysis of the gusset plate can be performed by both the Whitmore method (Ref. 15.2) and by evaluating the block shear mode of failure (Refs. 15.7 and 15.9). The more severe requirements resulting from these examinations should then be applied.

The Whitmore method required the evaluation of an effective plate area, as indicated in Fig. 15.4. The normal stress on this effective area should not exceed that permitted by the governing specification. The block shear strength is obtained by calculating the shear resistance obtained on the two possible failure surfaces along the length of the joint and adding the tensile resistance obtained across the end. For short joints, these areas should be taken through the bolt holes. For long joints, the gross areas can be used.

REFERENCES

15.1 E. H. Gaylord and C. N. Gaylord, *Design of Steel Structures*, 2nd ed., McGraw-Hill, New York, 1972.

15.2 R. E. Whitmore, *Experimental Investigation of Stresses in Gusset Plates*, University of Tennessee Engineering Experiment Station Bulletin 16, May 1952.

15.3 P. C. Birkemoe, R. A. Eubanks and W. H. Munse, *Distribution of Stresses and Partition of Loads in Gusseted Connections*, Structural Research Series Report 343, Department of Civil Engineering, University of Illinois, Urbana, March 1969.

15.4 D. D. Vasarhelyi, "Tests of Gusset Plate Models," *Journal of the Structural Division, ASCE*, Vol. 97, ST2, February 1971.

15.5 C. S. Davis, "Computer Analysis of the Stresses in a Gusset Plate," M.S. Thesis, Department of Civil Engineering, University of Washington, Seattle, 1967.

15.6 J. H. A. Struik, "Applications of Finite Element Analysis to Non-Linear Plane Stress Problems," Ph.D. Dissertation, Department of Civil Engineering, Lehigh University, Bethlehem, Pennsylvania, November 1972.

15.7 R. M. Richard, D. A. Rabern, D. E. Hormby, and G. C. Williams, "Analytical Models for Steel Connections," *Behavior of Metal Structures, Proceedings of the W. H. Munse Symposium*, Edited by W. J. Hall and M. P. Gaus, ASCE, May 17, 1983.

15.8 S. K. Chakrabarti and R. Bjorhovde, Tests of Gusset Plate Connections, Department of Civil Engineering, University of Arizona, Tucson, April 1983.

15.9 R. M. Richard, "Analysis of Large Bracing Connection Designs for Heavy Construction," Proceedings, National Engineering Conference, American Institute of Steel Construction, Chicago, 1986.

15.10 J. A. Waddell, *Bridge Engineering*, Wiley, New York, 1916.

15.11 T. H. Rust, "Specification and Design of Steel Gusset Plates," *Transactions, ASCE*, Vol. 105, 1940.

Chapter Sixteen

Beam and Girder Splices

16.1 INTRODUCTION

Splices in beams and girders are generally classified either as shop or field splices. Shop splices are made during the fabrication of the member in the shop. They are usually required to overcome length limitations of structural components as a result of fabrication or transportation facilities. The location of a shop splice in a member is often determined by loading conditions or stress resultants acting on the member and by the available lengths of material. Splices may also be introduced to permit the size of the cross-section to be changed with length in order to meet strength requirements more closely.

Field splices are necessary when a structural member becomes too long to be transported in one piece from the shop to the construction site. Occasionally, the available equipment in the field may also limit the maximum size or weight of structural components. Such limitations may require additional field splices.

This chapter deals specifically with the analysis and design of bolted or riveted beam and girder splices. Current practice varies and is largely based on past experience and limited experiment data.[16.1–16.3] Most designs involve equilibrium checks of the joint components using either allowable stress design or load factor design, as appropriate. Past practice has shown that this procedure results in a satisfactory design when the connection is subjected to static loading. Further work may lead to the development of more rational methods of analysis for this type of splice.

16.2 TYPES AND BEHAVIOR OF BEAM OR GIRDER SPLICES

Two types of connections are currently in use for bolted beam or girder splices. They are (1) the end-plate connection and (2) the more commonly used web-flange splice. Both connections are shown in Fig. 16.1. The major difference between these two types of joints is the loading condition to which the fasteners are subjected. The fasteners in the end-plate connection are generally subjected to a combined axial force and shear force, whereas the fasteners in the web-flange-type splice are subjected to shear alone. The end-plate connection is also used as a

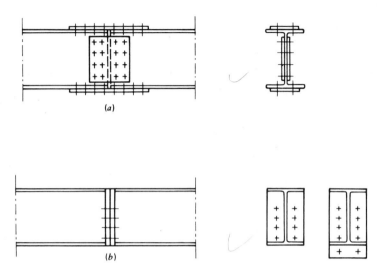

Fig. 16.1. Beam splices. (*a*) Web and flange splice. (*b*) End-plate splice.

moment-resistant beam-to-column connection. The design recommendations for end plates in beam-to-column joints discussed in Chapter 18 can also be applied to the end plate connection in a beam splice. After an initial discussion of the relative performance of the two types of connections, emphasis in this chapter is placed on the design of web-flange-type splices.

Usually two bolts are placed in the compression region of an end-plate connection. Although these bolts do not actively participate in transferring the moment, they are desirable from a practical point of view and serve to maintain the geometry of the joint. They also increase the shear capacity of the joint. In addition to the bolts in the compression region, a cluster of bolts is placed near the tension flange in order to obtain the maximum moment resistance for a given number of bolts and type of end-plate. The fasteners near the tension flange can be used even more effectively if the end-plate is extended beyond the tension flange and bolts are placed in this region as well (see Fig. 16.1).

As a moment connection, the end-plate splice is most economical in relatively light constructional steelwork because it requires less material and fasteners than conventional web-flange splices. Satisfactory behavior up to the plastic limit load of the beam can be achieved if the fasteners are adequately designed. This is illustrated in Fig. 16.2 where load versus midspan deflection curves are compared for beams with two types of end-plate splices in the constant moment region.[16.2] The observed behavior is almost identical to the behavior of plain beams. The plastic moment for the gross section of the beam was reached and sustained.

As beam sizes are increased or when large shear forces are to be transferred, the end-plate splice loses much of its economy and is replaced by the conventional

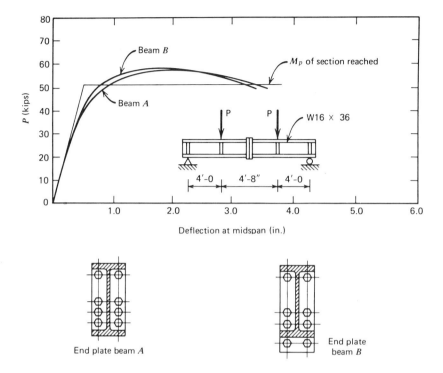

Fig. 16.2. Load versus deflection diagrams for beams with end-plate splice (Ref. 16.2).

beam splice shown in Fig. 16.1a. The location of the web and flange splices may be staggered, but this is often avoided to simplify field assembly.

In the design of girder splices, it is customary to assume that the web transmits the shear force and the flange splices resist the moment. The effect of these assumptions is examined in Subsections 16.2.1 and 16.2.2.

16.2.1 Flange Splices

Investigations were performed to determine the ultimate resisting moment of a beam with fastener holes in both flanges.[16.1–16.3] The general objective of these investigations was to evaluate whether the plastic moment capacity of the gross cross-section could be developed and whether the connection could provide sufficient rotation capacity. An extensive test series was reported in Ref. 16.2. Plain beams, beams with holes in the flanges, and beams with a flange splice in the constant moment region were tested. Single splice plates were bolted on the outside of the flanges and the allowable fastener shear varied from 15 to 30 ksi. Typical results are shown in Fig. 16.3. The nondimensional load versus deflection curves show the ratio of load to first yield load and deflection to yield load deflection. Figure 16.3a compares the behavior of a plain rolled beam to that of a beam with

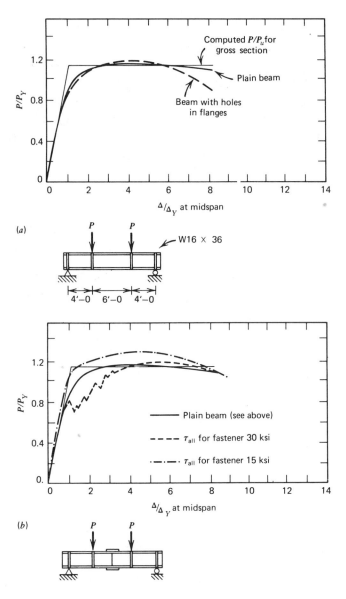

Fig. 16.3. Typical load versus deflection curves (Ref. 16.2).

holes in the flanges. No splice plates were provided. It is apparent that the holes did not affect the flexural capacity of this beam. Figure 16.3*b* shows the load versus deformation behavior of similar beams with the flanges spliced. The required number of fasteners for the splice was based on an allowable shear stress for the fasteners of 15 ksi for one beam, which resulted in 48 fasteners per splice. A second beam was designed using 30 ksi in shear which resulted in 24 fasteners per splice. An allowable shear stress of 15 ksi for clean mill scale surfaces is a conservative estimate of the capacity of a slip-resistant joint. Therefore, slip was not expected to develop in this joint and did not occur.

In all cases, the plastic moment capacity of the gross cross-section was developed, even though two $\frac{15}{16}$-in. diameter holes were placed in each flange cross-section. This reduced the flange area by 23%. Nevertheless, the beams were all able to develop the full plastic moment of the gross section: the holes in the flanges did not decrease the moment capacity of the beams.[16.2] The holes only influence the strain in the flanges locally. The material near the net section at the holes strain-hardened and permitted the full plastic moment of the gross section to be reached. This behavior of the net section is related to the ratio of the net to gross section area of the flanges, as was noted in Chapter 5.

Figure 16.3b shows that, although the slip between the splice plates and the flanges influences the load versus deformation behavior of the beam, it has a negligible effect on the ultimate moment capacity of the beam.[16.2, 16.3] At ultimate, plastic hinges formed in the constant moment region and failure occurred by local buckling of the compression flange.

In the beam tests reported in Ref. 16.2 flange splices were present in the constant moment region. There was no web splice present. As illustrated in Fig. 16.3b, in this situation where moment was present but shear was zero, properly proportioned flange plates alone were able to provide full moment transfer. This should be generally true for beams of usual proportions, that is, for beams in which the flange material constitutes the majority of the cross-section. This observation is further confirmed by tests reported in Ref. 16.1 in which both web and flange splice plates were used in a constant moment region.

The observed maximum moment capacity in these test beams was approximately equal to the gross section plastic moment. Hence, providing web splice plates did not significantly alter the moment capacity of the beam.

16.2.2 Web Splices

Figure 16.4a shows a splice made in a region of a beam in which both shear and moment are present. Because the transverse stiffness of flange splice plates will usually be very small, it must logically be assumed that the web splice has to carry all the transverse shear at the section. A free-body diagram through one set of fasteners is shown in Fig. 16.4b. For the ultimate strength of the fasteners in the web splice to be just reached, the deformation in bolts 1 and 3 must just attain their maximum shearing deformation. The corresponding ultimate forces, R_1 and R_3, are shown on the free-body diagram. Assuming rigid-body rotation of the connected parts, the directions of R_1 and R_3 will be perpendicular to a radius from the instantaneous center of rotation. The deformations of any other fasteners, only R_2 in this illustrative example, will be proportional to their distance from the instantaneous center of rotation. The corresponding fastener forces for these other bolts can be established from the load versus deformation response of the bolts acting in shear, that is, from an expression like Eq. 13.8.

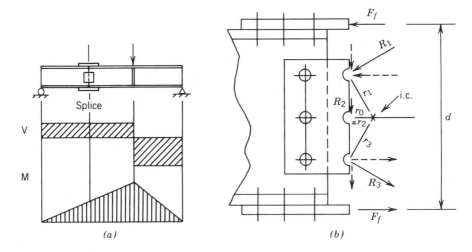

Fig. 16.4. Analytical model for web splice.

Calling the shear and moment at the section V and M, respectively, the equations of equilibrium can be written as

$$\sum_{i=1}^{n} R_{iv} - V = 0 \tag{16.1}$$

$$\sum_{i=1}^{n} R_i r_i + F_f d - M = 0 \tag{16.2}$$

Equation 16.1 says that the transverse shear at the section is resisted by the vertical components of the bolt forces. Equation 16.2 identifies how the moment at the section is shared between the flange splice plates and the web splice plates. Note that if there is moment at the section but no shear, the instantaneous center of rotation will be at the center of gravity of the fastener group being examined (in this case, the right-hand side three bolts). If there is shear at the section but no moment, as can occur in a continuous beam, the instantaneous center of rotation will be at infinity (to the right-hand side) from the cut section. In general, however, the location of the instantaneous center of rotation will have to be established, by trial, such that Eqs. 16.1 and 16.2 are satisfied.

As outlined earlier, the assumption that a properly proportioned flange splice can carry the full moment capacity of the cut section seems to give satisfactory results compared with experimental evidence. Of course, it will be a conservative solution if a web splice is also present because the web splice will also carry moment, in accordance with Eq. 16.2. Selection of the size and arrangement of fasteners in the web splice has not been subjected to the same experimental scrutiny, however. The use of Eqs. 16.1 and 16.2 can be applied on a case-by-case basis. Obviously, it would be advantageous from the point of view of design if the location

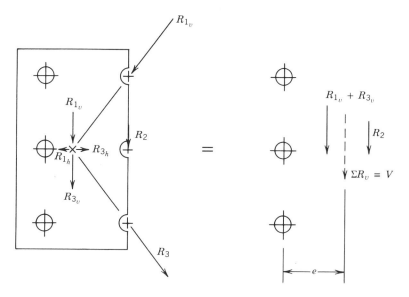

Fig. 16.5. Resolution of web splice forces.

of an eccentrically placed shear force that would yield results identical to that given by Eqs. 16.1 and 16.2 could be established. Figure 16.5 shows how the forces in the bolts on the right-hand side of the web splice of Fig. 16.4 can be resolved to locate the shear force acting on the bolts on the left-hand side of this splice. Unfortunately, the eccentricity "*e*" bears no particular relationship to the center of gravity of either fastener group. Indeed, it can lie on either side of the right-hand bolt group. It is not necessarily conservative, therefore, to assume that the fastener group on one side of the splice can be designed on the basis of a shear force acting through the center of gravity of the fastener group on the other side. Design on that basis does not seem to have led to difficulties in the past, however. Experimental and analytical studies of this problem are currently (1987) underway.

16.3 DESIGN RECOMMENDATIONS

16.3.1 Flange Splices

The flange splice can be designed conservatively by assuming that it transfers all the moment at the section. The fasteners in each flange must resist the force in the flange, taken as equal to approximately the moment at the cross-section divided by the beam depth (M/d). A single shear splice plate on each flange is often sufficient. For large shapes and for heavy flanges, splice plates may be required on both sides of the flanges in order to reduce the number of fasteners by providing a double shear condition and to reduce the splice plate thickness. The fasteners can be designed using the recommendations given in Chapter 5 for symmetric butt joints. Depending on the required joint performance, either slip-resistant or bearing-type joints can be used.

The moment capacity of the beam is not affected by the reduction in cross-sectional area caused by the fastener holes unless the ratio of net section to gross section area of the flanges (the A_n/A_g ratio) is less than $\sigma_y/0.85\sigma_u$ (see Chapter 5). The flange splice plates in the tension region should be treated as tension members and are also subject to the design recommendations given in Chapter 5.

16.3.2 Web Splices

The fasteners in the web splice should be designed such that the vertical components of the bolt forces are sufficient to carry the transverse shear at the section. In load factor design, this can be accomplished by meeting the requirements of Eqs. 16.1 and 16.2. In allowable stress design, the same examination can be made (that is, use Eqs. 16.1 and 16.2) and the resulting capacity then reduced in the ratio of permissible load for a single bolt in load factor design to that of a single bolt in allowable stress design. The web splice can also be designed as slip resistant. In this case, the same general procedure is to be followed except that all bolt forces are taken as equal (see Subsection 13.3.1).

As an alternative to the above, the current procedure for the design of web splices can be followed. In this case, design the bolt group on one side of the web splice for an eccentric force (equal to the shear at the section) acting through the center of gravity of the bolt group on the other side of the splice. The procedures and design recommendations given in Chapter 13 for eccentrically loaded connections can then be employed. Although this procedure has resulted in splices that have given satisfactory performances in the past, the actual margin of safety is unknown.

Two web splice plates, one on either side of the web, are recommended for beam or girder splices. This not only creates a symmetric load transfer with respect to the plane of the web, but also produces double shear conditions and thereby reduces the required number of fasteners and thus the eccentricity.

The overall dimensions of the web splice plates depend on the selected fastener pattern. The thickness of the splice plate can be determined from the applied eccentric shear load and the applicable shear, bending, and bearing stresses.

The fastener shear stresses and the bearing stresses suggested in Chapter 5 were shown in Ref. 18.7 to be fully applicable.

REFERENCES

16.1 F. W. Schutz, Jr., "Strength of Moment Connections Using High Tensile Strength Bolts," *AISC National Engineering Conference, Proceedings*, 1959.

16.2 R. T. Douty, and W. McGuire, "High Strength Bolted Moment Connections," *Journal of the Structural Division, ASCE*, Vol. 91, ST2, April 1965.

16.3 L. G. Johnson, J. C. Cannon, and L. A. Spooner, "Joints in High Tensile Preloaded Bolts—Tests on Joints Designed to Develop Full Plastic Moments on Connected Members," Jubilee Symposium on High Strength Bolts, Institution of Structural Engineers, London, June 1959.

Chapter Seventeen

Tension-Type Connections

17.1 INTRODUCTION

Fasteners are often subjected to a tensile-type loading by T-stubs or their equivalent. Some typical examples in this category are the hanger connection, the diagonal brace connection, and the beam-to-column connections shown in Fig. 17.1. Depending on the direction of the bending moment, either the top or bottom flange T-stub in a beam-to-column connection (Fig. 17.1a) is stressed in tension. It has long been recognized that deformation of the T-stub can result in additional fastener tension.[17.1] This phenomenon is called prying action. Tests have indicated that prying action can reduce both the ultimate load capacity and the fatigue strength of bolted and riveted joints.[16.2, 17.1–17.4]

17.2 SINGLE FASTENERS IN TENSION

Cooling of hot-driven rivets or tightening of a nut on a bolt results in an axial force or preload in the fastener. Inasmuch as this load exists prior to the application of external loading, the fastener is prestressed. As a result of this preload, the externally applied loads mainly change the contact pressure between the plates; very little additional fastener elongation is introduced and therefore there is only a minor change in bolt tension. This behavior can be illustrated by the model shown in Fig. 17.2.[13.11, 17.7] Tightening of the nut results in a tension in the bolt and compression between the connected parts. Assuming that the bolts and plates remain elastic, the force in each is proportional to its change in length, that is

$$\Delta B = k_b \Delta e \qquad (17.1)$$

and

$$\Delta C = -k_p \Delta e \qquad (17.2)$$

where B represents the bolt preload, C is the summation of contact forces between the plates and k_b and k_p the stiffness of the bolt and the gripped plates, respectively.

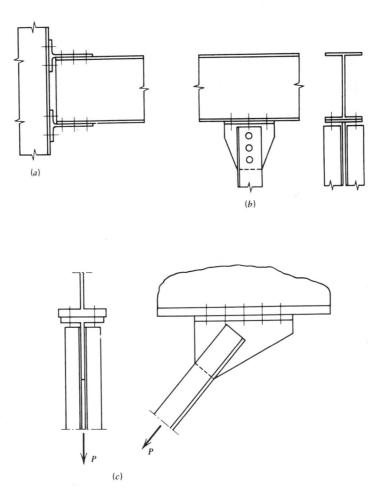

Fig. 17.1. Typical uses of T-type structural connections. (*a*) Beam-to-column connection. (*b*) Hanger connection. (*c*) Diagonal brace connection.

The term Δe represents the change in bolt elongation due to an externally applied load. As long as separation of the plates does not occur, the change in bolt elongation is equal to the change in thickness of the precompressed parts.

For the usual bolt and plate combinations, k_p will be much larger than k_b because the force B_0 is concentrated in the bolt whereas the force C_i is distributed over a much larger area, the effective contact area of the plates. If no load is applied to the connection, the bolt preload B_0 and the contact forces C_i are equal (Fig. 17.2a). When a load T is applied to the connected parts (the plate in Fig. 17.2b), the fastener will elongate and the precompressed plates tend to expand. If the expansion does not exceed the initial contraction of the plates, some contact pressure will remain (Fig. 17.2b). Now, the requirement of equilibrium can be stated as

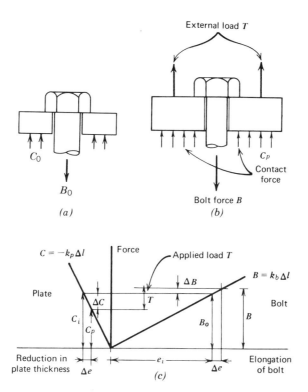

Fig. 17.2. Force in prestressed fastener.

$$B = C_p + T \tag{17.3}$$

where T is the externally applied load, C_p the summation of the reduced contact forces, and B the bolt force under an applied force T. Under such conditions an increase in applied load T results in an increase in bolt elongation Δe. For compatability, plates must expand by the same amount. As illustrated in Fig. 17.2c, because of the differences in bolt and plate stiffness, the application of the external force T results in a greater change in the compression in the plates (depicted as ΔC) than in the tension in the bolt, indicated as ΔB. Further increases in the external load T eventually reduce the contact pressure between the plates to zero, and the parts are on the verge of separation. For elastic conditions, separation of the plates takes place at an applied load equal to

$$T = B_0 \left[1 + \frac{k_b}{k_p} \right] \tag{17.4}$$

After the plates are separated, the bolt force B is simply equal to the external applied load T.

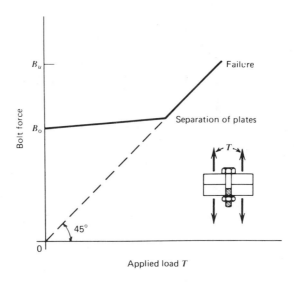

Fig. 17.3.   Bolt force versus applied load for prestressed single bolt connection.

The complete variation of the bolt force as a function of the applied load is given diagramatically in Fig. 17.3. The factor k_b/k_p depends on actual dimensions of the connection. However, for most practical cases the ratio varies between 0.05 and 0.10.[13.11] Hence, unless separation of the plates takes place, the maximum increase in bolt force due to an applied external load is of the order of 5 to 10% of the initial bolt preload.

17.3 BOLT GROUPS LOADED IN TENSION—PRYING ACTION

One of the simplest connections with the bolt groups in tension is the symmetric T-stub hanger with a single line of fasteners parallel to and on each side of the web. Because of symmetry of the connection, the fasteners can be assumed to be stressed equally. An external tensile load on the connection will reduce the contact pressure between the T-stub flange and the base. However, depending on the flexural rigidity of the T-stub, additional forces may be developed near the flange tip. This phenomenon is referred to as prying action and is illustrated in Fig. 17.4. The prying action increases the fastener force and this increase must generally be taken into account in the analysis of the connection.

The idealizations used to evaluate prying forces have the effect of only increasing the axial force in the bolt. In fact, the distortion of the connected parts also results in bending of the bolt, and local bending of the bolt nut or head can be significant, even when there is no appreciable increase in bolt axial force.

If the flange of a T-stub connection is sufficiently stiff, the flexural deformations of the flange will be small compared with the elongation of the fasteners. Very

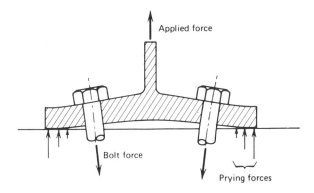

Fig. 17.4. Schematic of joint deformation.

little prying force will be developed, and the fastener will behave much like a single bolt in tension. This is illustrated in Fig. 17.5a where the bolt force in a test specimen is plotted as a function of the external applied load. The maximum moment in the T-stub occurs at the interface between the web and the flange. Since very little prying force is developed, the flange is subjected to single curvature bending.

When more flexible T-stub flanges are used, the flexural deformation of the flange induces prying forces that result in the additional bolt forces illustrated in Fig. 17.5b. Initially, the external load reduces the contact pressure between the flange and the base until separation at the bolt line occurs. Bending in the outer portions of the flanges develops prying forces acting between the bolt line and the edge of the flange, as illustrated in Fig. 17.4. Yielding of the fasteners and the T-stub flange often permits an increase in the applied load with only a small increase in bolt force. Because of this plastic flow, the prying force is reduced at this load level (see Fig. 17.5b). Depending on the flexural rigidity of the flange and the properties of the fasteners, prying forces may persist up to the point of failure.

Test results have confirmed that the stiffness properties of both the flange and the fasteners are significant factors influencing the prying action.[16.2, 17.2–17.4] Other factors, such as the magnitude of the initial clamping force of the fasteners, the grip length, and the number of lines of fasteners, have also been studied. Test results have indicated that the initial clamping force does not affect the prying action at ultimate load.[17.2, 17.3] This is illustrated in Fig. 17.6 where joints with two different bolt preloads are compared. The bolt force in the T-stub connection is plotted as a function of the applied load. The prying action at load levels close to the ultimate load was about the same for both conditions.

Although an increase in grip length may reduce the prying action at relatively low loads, the behavior at ultimate load is not significantly affected.[17.2, 17.3] The prying action at ultimate load is influenced by the deformation capacity of the bolts. At ultimate load, the inelastic deformations of the threaded portion of the bolt are

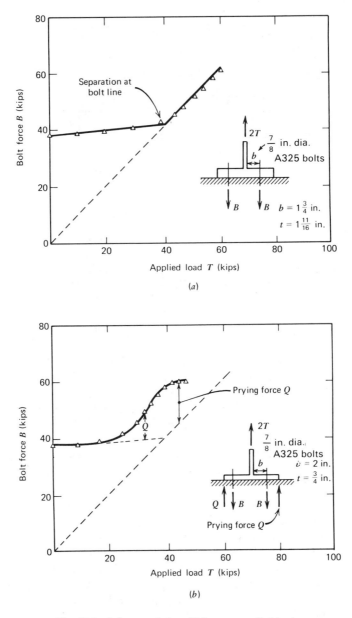

Fig. 17.5. Influence of plate thickness on applied load.

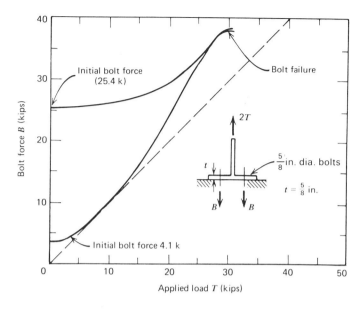

Fig. 17.6. Influence of initial bolt preload on prying action (Ref. 17.3).

more critical than the small elastic elongations that occur in the bolt shank. An increase in grip length has only a minor effect as long as the length of the thread under the nut is relatively constant.

In the discussion so far, it has been assumed tacitly that the T-section is connected to a rigid base. However, practical situations do arise wherein the member to which the T-section is connected does not provide a rigid base. A typical example is a T-section that transfers the tensile component in a moment resistant beam-to-column connection. The web of the T-stub is connected to the beam tension flange, and the flange of the T-section is bolted to the column flanges (see Fig. 17.1). If the column flanges do not provide adequate stiffness under the applied load system, the location of the prying forces may shift from the toe lines *AB* and *CD*, to the edges *AC* and *BD* (see Fig. 17.7*a*). In such connections the magnitude and the location of the prying forces are governed by the relative stiffness of the T-stub flange and the column flange. Generally, the resulting loading condition in such a connection is highly complex and has not been studied extensively. Reference 17.5 summarizes the results of a series of tests in which T-sections were bolted to the flanges of a wide flange shape. The T-sections were loaded in tension. The influence of the column flange thickness on the location and the magnitude of the prying forces was studied[17.5] and some typical test results are shown in Fig. 17.7*b*. It is apparent from the deformation pattern that as the stiffness of the T-flange is increased, the prying forces tend to concentrate in the areas near the corners of the T-section. When the stiffness of the T-stub flange is much greater than the stiffness

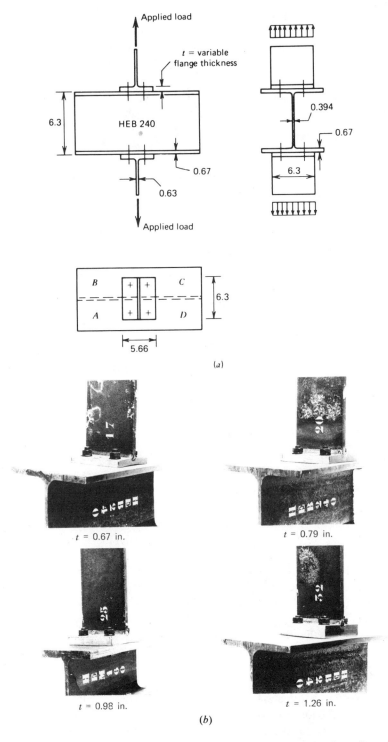

Fig. 17.7. T-stub sections bolted onto nonrigid support (Ref. 17.3). (*a*) Specimen dimensions. (*b*) Deformation pattern for various T-stub flange stiffnesses. (Courtesy of Stevin Laboratory, Delft University of Technology.)

of the column flange, the T-section provides the rigid base, and prying forces are developed because of deformations of the column flange.

Bouwman has conducted tests in which the contact surface between pairs of opposing T-stubs was uniquely established.[17.8] The two extremes are shown in Fig. 17.8. Fig. 17.8a shows a specimen in which the contact points are located at the extremities of the T-stub flange. Clearly, this represents a case in which a great deal of prying action can be present, depending upon the flexibility of the flange. Static tests of this arrangement showed that prying forces were present from the beginning of loading (for the particular geometry involved), and the prying force at the time of separation of the parts was about equal to the magnitude of the applied load. The arrangement shown in Fig. 17.8b will result in relatively little prying force in the fasteners if the area of contact is small and the flange is relatively stiff or the bolts relatively close to the T-stub web. For the dimensions used in the test pieces (32-mm flange thickness and 20-mm dia. bolts 70 mm from the centerline of the web), there was virtually no increase in bolt force until separation of the parts. However, with a more flexible flange, bending of the bolt will become more significant, particularly at the junction of the bolt head and the shank. Appreciable "prying" forces can develop, even with the arrangement shown in Fig. 17.8b.

When hangers have more than two rows of fasteners parallel to the web (see Fig. 17.9a), the effectiveness of the outer rows may be sharply reduced because of the flange flexibility. Tests have demonstrated that, upon loading of the connection, the strain in the inner fasteners increased and continued to do so until failure occurred.[17.2] However, initially the strain in the outer bolts decreased slightly or remained constant. Thus, in the early stages of loading, almost the entire load is carried by the inner bolts. Failure of the inner fasteners occurred before the strength of the outer fasteners could be developed. Increasing the flexural stiffness of the flange resulted in increased efficiencies. Test efficiencies between 45 and 80% were observed.[17.2] This shows that the outer bolts are not very effective in carrying the applied load unless the flanges are extremely heavy or stiffened, as indicated in Fig. 17.9b. If stiffeners are used, their connection to the T-stub web

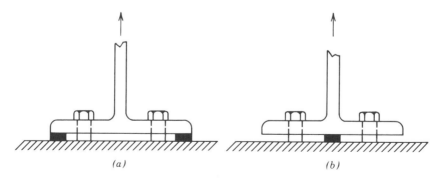

(a)   (b)

Fig. 17.8.   Test specimens used to establish influence of contact surface on prying.

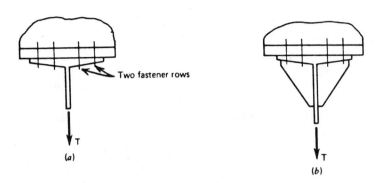

Fig. 17.9. Four-row hanger connections.

and flange must be properly proportioned. The stiffeners must function until the ultimate load of the connection is reached. If the connection fails prematurely, the now unstiffened T-stub can develop large prying forces, and bolt failure may occur at levels lower than expected.[17.9]

17.4 REPEATED LOADING OF TENSION-TYPE CONNECTIONS

As early as 1956 it was identified that prying forces could significantly reduce the fatigue strength of a tension-type T-connection.[17.1] Although extensive data are not available, further research has yielded information on the behavior of bolted T-connections under repeated loading conditions.[17.4, 17.8] Fatigue tests were carried out on connections having a single line of fasteners on either side of the web.

The bolt tension history of a single fastener installed in a plate assembly and subjected to an external tensile load was discussed earlier. The idealized relationship between the axial force in the bolt and the applied load is summarized in Fig. 17.3. The results plotted in Fig. 17.5a indicated that relatively stiff tension-type T-connections behave similarly to single bolt and plate assemblies. For such cases it is apparent that any increase in bolt force due to application of an external force will be small as long as the connected parts do not separate. This means that the stress range in the bolt will likewise be small, and the fatigue life of the bolts will be relatively long. This situation can be said to represent one extreme of the fatigue life behavior. It must be emphasized that the long fatigue life depends upon a combination of factors: negligible prying forces, little or no bending in the bolt threads, and a high level of initial preload in the bolt. Figure 17.10 shows the fatigue life of a T-stub connection that used A36 steel connected by $\frac{3}{4}$-in. diameter A490 bolts. Illustrated are three different levels of prying force. (In all cases, bolts were installed by the turn-of-nut method and had at least the minimum required preload at the time the cyclic loading was started.) The specimen with the smallest amount of prying had the longest fatigue life; it had still not broken at 3×10^6 cycles of loading, at which time testing was discontinued.

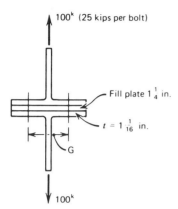

Fig. 17.10. Bolted T-stubs under repeated loading conditions (Ref. 17.4). Asterisk below denotes connection that did not fail. Test discontinued.

G (in.)	Applied Load Range per Bolt kips	Static Prying Ratio Q/T at Ultimate Load (Method of Ref. 17.4)	Range in Average Bolt Stress First Cycle (ksi)	Number of Cycles to Failure
3	0–25	0.02	2.2	3,000,000*
$4\frac{1}{2}$	0–25	0.19	3.7	592,000
6	0–25	0.45	10.4	32,000

If the flanges of a T-connection loaded in tension are flexible, prying forces develop and a significant decrease in fatigue life can result.[17.4] The data in Fig. 17.10 show clearly that the fatigue life of these connections decreased dramatically as the level of prying force increased. As the prying ratio decreased from 0.45 to 0.02 (as calculated in Ref. 17.4 and for conditions at ultimate load), the number of cycles to failure by fatigue increased by a factor of at least 100. Measured values of the average stress range in the bolts decreased from 10.4 to 2.2 ksi over this interval. Similar connections fastened with A325 bolts exhibited the same behavior, and the results have been generally confirmed by other researchers. It is apparent that an increase in prying force resulted in a decrease in fatigue life of the connection. These reductions can be explained qualitatively by examining the prying forces during a fatigue-type loading.

As illustrated in Fig. 17.5b, the prying force Q in a flexible connection resulted in a large increase in bolt load as compared with the relatively rigid connection shown in Fig. 17.5a. The more flexible connection results in a greater stress range in the fastener, with a corresponding decrease in fatigue strength. In addition, flexural deformations in the flange may distort the thread region of the bolt shaft. This also results in a higher stress range at the root compared with the average stress range in the bolt.

If the applied load on the connection is sufficient to produce a yielding of the fasteners, a reduced clamping force results upon unloading. Subsequent cycles of

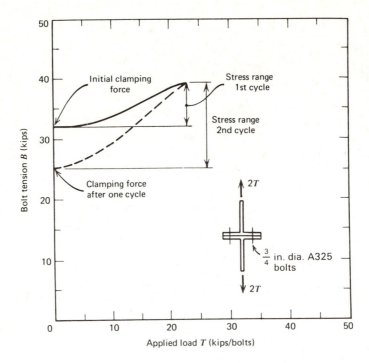

Fig. 17.11.   Influence of prying force on fastener clamping force after unloading (Ref. 17.4).

load result in an increase in stress range. This is shown in Fig. 17.11 for a carbon steel T-connection fastened by $\frac{3}{4}$-in. dia. A325 bolts.[17.6] An applied load of 24 kip/bolt increased the bolt load by about 7 kip. Upon unloading, the initial clamping force was reduced from 32 kips to about 25 kips. When the external load was reapplied, the stress range during the second cycle was almost twice the stress range observed during the first cycle. A static test of an identical connection yielded a prying ratio Q/T equal to 0.37 at ultimate load.[17.4] When the same external load (24 kips/fastener) was applied to a connection in which very little prying force was developed, the increase in bolt load was only about 2 kips and the initial clamping force was not noticeably reduced after unloading. Subsequent cycles yielded a similar bolt load change.

These studies illustrate that large prying forces not only decrease the static strength of the connection but also have a detrimental effect on the fatigue strength of the fasteners. It is apparent that a connection that develops little or no prying force is preferable under repeated loading.

17.5 ANALYSIS OF PRYING ACTION

Analytical and experimental studies of prying action have resulted in several mathematical models.[16.2, 17.3, 17.4, 17.6, 17.10] Douty and McGuire used the model shown in Fig. 17.12 and suggested a formula based on an elastic analysis. They considered the properties of the bolts and the connected material and the geometry of the connection. These formulas were then modified to simplify application and reflect test results. The following semi-empirical equation was obtained.

$$Q = \left\{ \frac{\frac{1}{2} - (wt^4/30ab^2A_b)}{a/b[(a/3b) + 1] + (wt^4/6ab^2A_b)} \right\} T \qquad (17.5)$$

This equation relates the prying force Q to the ultimate load of the connection. A similar formula with different coefficients was suggested for evaluating the prying force under working load conditions.[16.2]

The development of Eq. 17.5 by Douty and McGuire was based on the fulfillment of both equilibrium and compatibility conditions. The latter was obtained by assuming that the T-stub flange acts as a simple beam between its tips, acted upon by the bolt forces and the applied load. The flange deflection at the bolt line was equated to the axial deflection of the bolt and the flange expansion in the thickness direction due to reduction in contact pressure. For this compatibility statement to be valid, both the T-stub flange and the bolt must remain elastic and the flange must continue to act with a span of $2(a + b)$. Although the first condition may be met approximately under working load conditions, it is unlikely to be valid at the time that the ultimate load of the connection is reached. The second condition, which in effect says that the force Q acts uniquely at the flange tips, is not likely to be satisfied at any load level, although the effect of this inaccuracy cannot be determined.

Agerskov[17.10] has presented a development for the prediction of prying action that is similar in some major respects to that of Douty and McGuire. He also used both equilibrium and compatibility equations to develop a prediction of prying

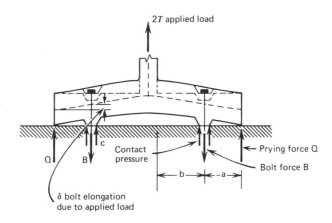

Fig. 17.12.   Model used by Douty and McGuire (Ref. 16.2).

forces. Agerskov provided a more sophisticated development of the bolt elongation and plate expansion terms. The possibility of washers is included and the bolt elongation is considered to be composed of nut, shaft, and threaded portion elongations. The possibility of inelastic deformations due to yielding of the flange is not included, however, and the span of the flange is again assumed to extend from flange tip to flange tip. Agerskov also identified the effect of shear as a reduction to the plastic moment capacity of the section. The normal stress due to bending and the shear stress are combined according to the von Mises criterion, and the plastic moment capacity is calculated using the reduced normal stress. It is generally accepted that such a reduction is quite conservative.[17.6] Rectangular sections of the proportions to be expected in the tension connections under consideration would undergo a theoretical reduction in plastic moment capacity resulting from shear of less than about 10%. Such a theoretical reduction would probably be masked by strain-hardening in an actual connection.

Because of its complexity, Eq. 17.5 is not readily suited for design. The semi-empirical relationship suggested by Douty and McGuire for the prying force at ultimate was simplified in Ref. 17.6 to yield (U.S. customary units)

$$\frac{Q}{T} = \left(\frac{3b}{8a} - \frac{t^3}{20}\right)$$

(17.6)

As is illustrated in Fig. 17.13, this equation tends to overestimate the prying force and provides conservative design results.[17.4]

An experimental and analytical study on connections consisting of two carbon steel T-stubs bolted together through the flanges with four A325 or A490 bolts was conducted at the University of Illinois and resulted in the development of empirical

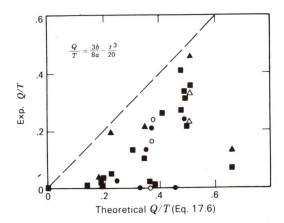

Fig. 17.13. Comparison between analytical and experimental results. ■ A325 bolts, $\sigma_{u\ spec}$ = 120 ksi. ▲ A490 bolts, $\sigma_{u\ spec}$ = 150–170 ksi. ● 10k bolts, $\sigma_{u\ spec}$ = 142 ksi. ○ 4D bolts, $\sigma_{u\ spec}$ = 50 ksi. △ A502 rivets, σ_u = 60–80 ksi.

formulas to approximate prying.[17.4] The prying ratio Q/T at ultimate load for connections with A325 bolts was given as

$$\frac{Q}{T} = \left(\frac{100bd^2 - 18wt^2}{70ad^2 + 21wt^2}\right) \tag{17.7}$$

For connections with A490 bolts, the coefficients 18 and 70 were replaced by 14 and 62, respectively. Use of Eq. 17.7 provided somewhat better agreement with the test results as compared with Eq. 17.6. Figure 17.14 shows the comparison between analytical and experimental results when Eq. 17.7 is used.

However, the empirical formulas are only applicable to the specific combination of bolt and plate material for which they were developed. Different formulas may be required for different bolt and plate material combinations.

A third analytical approach for predicting the prying force was suggested in Ref. 17.3. The simplified model, shown in Fig. 17.15, was used to describe the prying action in a T-stub with its flange bolted to a rigid base. The approach is not restricted to specific bolt and plate combinations since all major parameters that influence the prying action are included in the model. The symbol Q denotes the prying force per bolt at ultimate and this force is assumed to act as a line load at the edge of the flange. Test results have shown this to be a reasonable assumption for conditions near ultimate as long as the edge distance a is within certain limits. The ultimate tensile load of the fastener is B, and the corresponding applied load

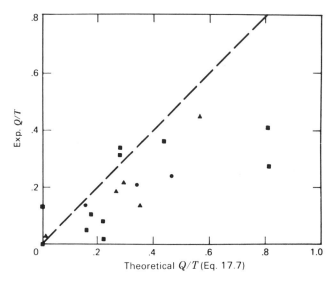

Fig. 17.14. Comparison between analytical and experimental results. ■ A325 bolts, ● 10k bolts: $Q/T = (100bd^2 - 18wt^2)/(70ad^2 + 21wt^2)$. ▲ A490 bolts, $Q/T = (100bd^2 - 14wt^2)/(62ad^2 + 21wt^2)$.

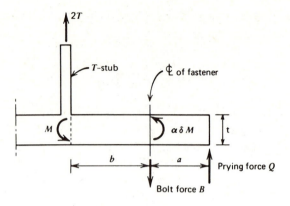

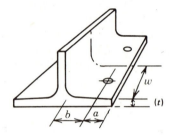

Fig. 17.15. Analytical model for prying force.

per bolt is equal to T. The bending moment at the interface between the web and the flange is taken as M, and the moment at the bolt line due to the prying force Q is taken to be equal to $\alpha\delta M$ where δ is equal to the ratio of the net area (at the bolt line) and the gross area (at the web face) of the flange. The term α represents the ratio between the moment per unit width at the centerline of the bolt line and the flange moment at the web face. Note that the factor α is a function of the unknown ratio Q/T. Using a free-body diagram of the portion of the T-stub flange between the bolt line and the web face, the equation of moment equilibrium gives

$$(1 + \delta\alpha)M = Tb \tag{17.8}$$

where b is the distance from the centerline of the bolt to the web. The ultimate moment capacity of the gross area of the flange is

$$M = \tfrac{1}{4}wt^2\sigma_y \tag{17.9}$$

where σ_y is the yield strength of the flange material, t is the flange thickness, and w the length of the flange parallel to the web that is tributary to each bolt (see Fig. 17.15).

Another equation can be obtained by writing the moment equilibrium requirement for the portion of the flange between the bolt line and the flange tip:

$$Qa = \alpha\delta\tfrac{1}{4}wt^2\sigma_y \qquad (17.10)$$

Equilibrium of applied load, bolt force, and prying force requires that

$$B = T + Q \qquad (17.10a)$$

When expressed in terms of the other moment and equilibrium conditions, this results in

$$B = T\left[1 + \frac{\delta\alpha}{(1 + \delta\alpha)}\frac{b}{a}\right] \qquad (17.11)$$

and

$$t = \left\{\frac{4Bab}{w\sigma_y[a + \alpha\delta(a + b)]}\right\}^{1/2} \qquad (17.12)$$

Equation 17.12 relates the required flange thickness to the mechanical properties and geometrical dimensions of the constituent parts of the connection. Experimental results and the prying ratio Q/T obtained from Eq. 17.11 are compared in Fig. 17.16 for different types of bolts. A few data obtained from riveted specimens are included as well.

It is apparent that the solution given by Eqs. 17.11 and 17.12 overestimates the prying force. The variation is comparable to that obtained when Eqs. 17.6 and 17.7 are used. Among the factors causing the difference between the load transfer predicted by the idealized model and the test results are strain-hardening and the actual distribution of forces. The model assumes the bolt force B to act at the centerline of the bolt. As a result of flexural deformations in the flange, the bolt force B is acting probably somewhere between the bolt axis and the edge of the bolt head, as indicated in Fig. 17.17. This decreases the distance b and changes the prying ratio Q/T directly. To approximate this effect, a revised equilibrium condition was developed using modified distances a' and b', defined in Fig. 17.18b. The resultant fastener force B was assumed to act at a distance b' equal to $b - d/2$ from the web face. The distance a' was taken equal to $a + d/2$. The model assumes the prying force Q at ultimate load to be a line load at the tip of the flange.

Tests have indicated that this is a reasonable assumption as long as the end distance is not much greater than the distance b. Therefore, it is recommended that the end distance a be limited to $1.25b$.

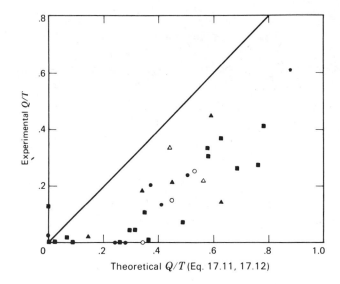

Fig. 17.16. Comparison between theoretical and experimental Q/T ratios. ■ A325 bolts, $\sigma_{u\ \text{spec}} = 120$ ksi. ▲ A490 bolts, $\sigma_{u\ \text{spec}} = 150$–$170$ ksi. ● 10k bolts, $\sigma_{u\ \text{spec}} = 142$ ksi. ○ 4.D bolts, $\sigma_{u\ \text{spec}} = 50$ ksi. △ A502 rivets, $\sigma_{u} = 60$–80 ksi.

As illustrated in Fig. 17.18*a*, the predicted prying force based on these modified dimensions provides much better agreement with the test results. Use of these modified dimensions is likely to result in a conservative design of the bolts, however, since the model still tends to overestimate the influence of the prying force for most cases.

Equations 17.11 and 17.12 (using the modified *a'* and *b'* values) are recommended in the following section for use in calculating the effect of prying forces.

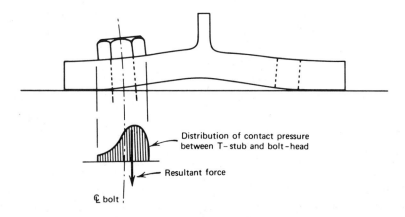

Fig. 17.17. Influence of flange deformations on location of resultant bolt force.

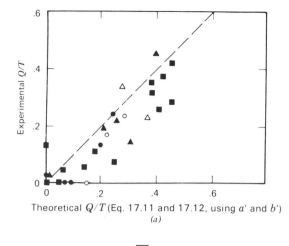

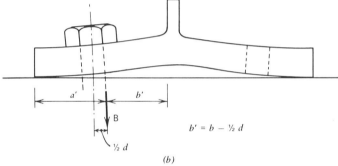

Fig. 17.18. Comparison between analytical and experimental Q/T ratios for modified a and b distances. ■ A325 bolts, $\sigma_{u \text{ spec}} = 120$ ksi. ▲ A490 bolts, $\sigma_{u \text{ spec}} = 150$–170 ksi. ● 10k bolts, $\sigma_{u \text{ spec}} = 142$ ksi. ○ 4.D bolts, $\sigma_{u \text{ spec}} = 50$ ksi. △ A502 rivets, $\sigma_u = 60$–80 ksi.

Designers must, however, be aware of the limitations of the formulae and be alert to details that are not comparable to those for which these equations have been developed. The equations give satisfactory predictions of failure for the range of connection geometries reflected by the test specimens. In a few tests, it has been observed that when the distance b is small relative to a and the flange is fairly flexible, there can be an exaggerated deformation of the bolt head or nut. Figure 17.19 illustrates this phenomenon. In these cases, the predicted prying force can be small, yet the bolt breaks at a load less than that associated with direct axial tension. In other words, there are really two phenomena related to prying action. The first in an overall prying (Fig. 17.4) that results in increased force in the bolt but one that is still generally axial. The second is a local prying of the fastener head (Fig. 17.19), producing both an axial force and bending in the bolt. Obviously, cases between these two extremes will also exist.

Although the problem of calculating the prying forces at ultimate load conditions

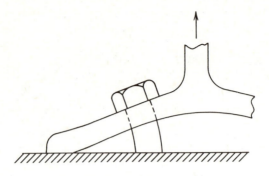

Fig. 17.19. Local prying of fastener head.

is complex, the solutions presented give reasonably satisfactory predictions of the test data. A prediction of the fatigue strength of the bolts in a tension connection is more complicated, and a satisfactory method of analysis is not yet (1987) available. The solution must include identification of the stess range in the bolt under service load conditions (including both direct prying force and bolt bending), recognition of the level of initial preload in the bolt, and calculation of local effects such as yielding in the vicinity of the bolt threads.

17.6   DESIGN RECOMMENDATIONS

17.6.1   Static Loading

Several semi-analytical and empirical approximations for the prying force in T-connections with a single line of fasteners on each side of the web have been examined. All of the methods provided about the same degree of fit to the test data.

A modification of the equilibrium method proposed by Struik and deBack[17.3] was observed to have several advantages. Of primary importance was the fact that it was applicable to a range of fasteners and steels and is readily suited for design. The analytical model used by Douty and McGuire had several coefficients adjusted on the basis of experimental work[16.2] and therefore it was not directly applicable to a variety of fasteners and materials. Similarly, the empirical formulas developed by Nair *et al.* were only applicable to specific plate and bolt combinations. Agerskov used a method similar to that of Douty and McGuire but used a compatibility requirement based on elastic conditions in combination with the ultimate strength of the cross-section.

Although the number of tests is limited, the experimental data are in reasonable agreement with predictions made using Eqs. 17.11 and 17.12, as illustrated in Fig. 17.18. The simplified model provides a satisfactory basis for designing bolted and riveted tension-type T-connections under static loading and for connections whose geometry is reasonably similar to those from which the test data were obtained.

Connection geometry that might result in severe local distortion of the bolt head may lead to ultimate loads lower than those predicted by Eqs. 17.11 and 17.12.

Connections with more than two gage lines of fasteners are not effective unless special provisions such as additional stiffening of the flange is provided.[17.2] If this is not provided, the load capacity is provided largely by the inner fasteners alone. It would be prudent to design stiffeners and their connections conservatively so that they will still be functioning at the time the ultimate strength of the bolts is reached. Premature failure of stiffeners would likely produce very high bolt prying forces.

i. Allowable Stress Design. The minimum tensile capacity of a fastener is equal to the product of the fastener stress area A_s and its minimum specified tensile strength $\sigma_{u\,spec}$ in kilopounds per square inch. As noted in Chapter 4, the tensile capacity of a bolt can be expressed in terms of the nominal bolt area A_b as

$$B_u = 0.75 A_b \sigma_{u\,spec} \tag{17.13}$$

Applying a factor of safety with respect to ultimate load equal to 2.0 yields an allowable tensile load B_{all} per fastener as

$$B_{all} = (0.5)(0.75) A_b \sigma_{u\,spec} \tag{17.14}$$

or

$$B_{all} = 0.375 A_b \sigma_u \, spec \tag{17.15}$$

A factor of safety of 2.0 is consistent with previously used values. It is also compatible with allowable shear and bearing stresses for bolts.

To provide a uniform margin between working load and ultimate strength, the applied load and prying force should not exceed the allowable bolt load. Hence

$$B_{all} \geq T + Q \tag{17.16}$$

The prying force Q depends on the geometrical dimensions of the connection, as well as upon the applied load T. These factors determine the value of α which is in turn related to the prying force Q, as given in Eq. 17.10. The design recommendations, summarized hereafter, can be used either for analysis or for design purposes.

The dimensions of the T-stub and the size, number, and location of the bolts all affect the strength of the connection. As such, it is usually expeditious to select tentative dimensions, sizes, and so on and then to analyze the connection. Adjustments can then be made if any part (T-stub or bolts) is either inadequate or overdesigned. Knowing the permissible fastener load, Eq. 17.11 can be used as an equality to solve for α, the moment ratio. (Following the nomenclature of Fig. 17.5, the total externally applied load on a two-bolt connection is described as 2T.) Values of α less than unity indicate that fastener capacity will control the design and Eq. 17.12 can then be used to determine whether or not the trial flange thickness is adequate. The value of B in Eq. 17.12 is to be taken as the permissible fastener load.

If the value of α determined using Eq. 17.11 exceeds unity, this indicates that the limiting condition has been reached. For this case, the ultimate load of the connection would be attained when plastic hinges form at the bolt lines and at the web-to-flange junction. In this situation, the force in the bolts at working load levels can be determined directly from Eq. 17.11 using $\alpha = 1.0$. In order to check the plate thickness (Eq. 17.12), the force in the bolts should be that corresponding to ultimate load levels. Hence, B is now to be taken as 2.0 times that established using Eq. 17.11.

It was noted in Section 17.5 that better agreement between test results and predictions made using Eqs. 17.11 and 17.12 was obtained if the distances a and b are modified as indicated in Fig. 17.18. Thus, for convenience, Eqs. 17.11 and 17.12 will be restated in the Design Recommendations For T-Connections Under Static Loading Conditions using the equality form and the modified geometry parameters.

DESIGN RECOMMENDATIONS FOR T-CONNECTIONS UNDER STATIC LOADING CONDITIONS

Allowable Stress Design

Allowable tensile load per fastener

$$B_{\text{all}} = 0.375 A_b \sigma_{u\,\text{spec}}$$

Check adequacy of fastener to resist the applied load and prying action:

$$B_{\text{all}} \geq T + Q$$

or upon substituting Eq. 17.11 with modified a and b distances;

$$B_{\text{all}} \geq T\left[1 + \frac{\delta\alpha}{(1 + \delta\alpha)}\frac{b'}{a'}\right] \tag{17.17}$$

The T-flange thickness must be equal to or exceed

$$t = \left\{\frac{4\bar{B}a'b'}{w\sigma_y[a' + \delta\alpha(a' + b')]}\right\}^{1/2} \tag{17.18}$$

where $a' = a + d/2$
$\quad\quad b' = b - d/2$
$\quad\quad \bar{B}$ = estimated fastener load at failure of the connection

if $a < 1.0$

$$\overline{B} = 0.75 A_b \sigma_u$$

if $\alpha \geqq 1.0$, it is taken as 1.0 and

$$\overline{B} = 2T\left[1 + \frac{\delta}{(1 + \delta)}\frac{b'}{a'}\right]$$

Maximum value of distance a

$$a \leq 1.25b$$

The design recommendations given in this section are valid for tension-type connections fastened to a rigid base. It was noted in Section 17.3 that the stiffness of the base to which the T-section is connected is an important parameter in the development of prying forces. If the base does not provide enough stiffness, the fastener loads and prying forces should be evaluated on the basis of the geometrical dimensions and material properties of the flange to which the T is connected. The joint component that provides the least stiffness results in the greatest prying forces and governs the design of the fasteners.

ii. Load Factor Design. The design of T-connections by load factor design is directly comparable to allowable stress design. The only difference is that the load on the fastener at the factored load level should not exceed the ultimate tensile load of the fastener multiplied by a reduction factor ϕ. A reduction factor ϕ equal to 0.85 is in reasonable agreement with past practice. A load factor of 1.7 and a reduction factor of 0.85 yields a design that is comparable to allowable stress design.

DESIGN RECOMMENDATIONS FOR T-CONNECTIONS UNDER STATIC LOADING CONDITIONS

Load Factor Design

Maximum tensile capacity of fastener

$$B = 0.75 A_b \sigma_u$$

Check adequacy of fastener to resist the applied load as well as prying action

$$\phi B \geq T'\left[1 + \frac{\delta\alpha}{(1 + \delta\alpha)}\frac{b'}{a'}\right]$$

where the reduction factor $\phi = 0.85$ and T' represents the applied load per fastener at the factored level. The T-flange thickness is given by

$$t = \left\{\frac{4\overline{B}a'b'}{w\sigma_y[a' + \delta\alpha(a' + b')]}\right\}^{1/2}$$

where $a' = a + d/2$
$\qquad b' = b - d/2$
$\qquad \overline{B}$ = estimated fastener load at failure of the connection

if $\alpha < 1.0$

$$\overline{B} = B = 0.75 A_b \sigma_u$$

if $\alpha \geqq 1.0$, it is taken as 1.0 and

$$\overline{B} = T' \left[1 + \frac{\delta}{(1 + \delta)} \frac{b'}{a'} \right]$$

Maximum value of distance a

$$a \leq 1.25b$$

17.6.2 Repeated Loading

The fatigue strength of high-strength bolts loaded in tension in T-connections is significantly affected by the preload in the fastener and by the amount of prying action present. Therefore, in situations where repeated loading is expected, special attention must be directed to bolt installation procedures to ensure that the bolts are properly tightened and provide the desired clamping force.

As noted earlier, prying forces in T-connections can lead to severe reductions in fatigue strengths. To avoid a reduction in strength and substantial decreases in life, the T-connection should be dimensioned so that prying forces are minimized. This can be accomplished by providing a reasonably rigid T-connection, as was shown in Fig. 17.5a. This will ensure that the applied load does not cause separation of the plates and that prying forces remain low. Consequently, the bolt will only experience a small change in stress as external load is applied.

Figure 17.20 shows the fatigue test results that are available.[17.4, 17.8] They represent two different bolt grades: A325 and A490 (or grade 10.9). In all cases, the bolts were identified as having a preload at least equal to the specified minimum value. The stress range parameter plotted in Fig. 17.20 corresponds to the bolt force calculated according to Eqs. 17.17 and 17.18 divided by the nominal area of the bolt. Although this appears to give a reasonable indication of the fatigue lives and is convenient for design, what should be calculated is the actual range of stress to which the bolts are subjected. This is much less than the nominal value used in Fig. 17.20. As has already been noted, the actual range of stress will depend on the stiffness of the parts (reflecting the amount of prying present) and the initial preload in the bolts. Measured stress range values are available for 12 of the test results shown in Fig. 17.20, and they are from about 5 to 25 times less than the values used to plot the figure.

In assessing the results shown in Fig. 17.20, it should also be noted that in only

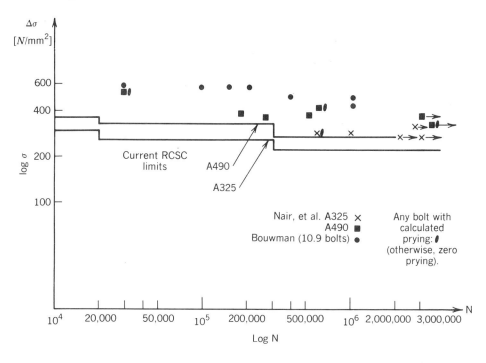

Fig. 17.20. Fatigue strength of bolts in T-stub connections.

four of the tests was there any calculated prying force, and even in these cases the prying forces were small.

The design recommendation given below has been based on the data shown in Fig. 17.20 (The recommendation is also shown in Fig. 17.20.)

DESIGN RECOMMENDATIONS FOR HIGH-STRENGTH BOLTS IN T-CONNECTIONS, REPEATED-TYPE LOADING

Permissible tensile forces in the bolts of T-stub connections shall not exceed

B_{all}	for number of cycles less than 20,000
$0.90\,B_{all}$	for number of cycles between 20,000 and 500,000
$0.75\,B_{all}$	for number of cycles greater than 500,000

The forces in the bolts shall be calculated under the specified loads using Eqs. 17.17 and 17.18. In no case shall the calculated prying force exceed 60% of the externally applied load.

REFERENCES

17.1 W. H. Munse, "Research on Bolted Connections," *Transactions, ASCE*, Vol. 121, 1956, p. 1255.

17.2 W. H. Munse, K. S. Peterson, and E. Chesson, Jr., "Strength of Rivets and Bolts in Tension," *Journal of the Structural Division, ASCE*, Vol. 85, ST3, March 1959.

17.3 J. H. A. Struik and J. de Back, *Tests on Bolted T-Stubs with Respect to a Bolted Beam-to-Column Connections*, Report 6-69-13, Stevin Laboratory, Delft University of Technology, Delft, the Netherlands, 1969.

17.4 R. S. Nair, P. C. Birkemoe, and W. H. Munse, "High Strength Bolts Subjected to Tension and Prying," *Journal of the Structural Division, ASCE*, Vol. 100, ST2, February 1974.

17.5 J. de Back and P. Zoetemeyer, *High Strength Bolted Beam-to-Column Connections, The Computation of Bolts, T-Stub Flanges and Column Flanges*, Report 6-72-13, Stevin Laboratory, Delft University of Technology, Delft, the Netherlands, 1972.

17.6 ASCE, *Commentary on Plastic Design*, Manual 41, New York, 1971.

17.7 J. L. Rumpf, "Riveted and Bolted Connections," in *Structural Steel Design*, 2nd Ed. Ronald Press, 1974, Chap. 18.

17.8 L. P. Bouwman, *Fatigue of Bolted Connections and Bolts Loaded in Tension*, Report No. 6-79-9, Stevin Laboratory, Delft University of Technology, Delft, the Netherlands, 1979.

17.9 James W. Baldwin, Jr., *Test Report, Kemper Arena Roof Hangers*, prepared for City of Kansas City Director of Public Works, November 12, 1980, Kansas City, Missouri.

17.10 H. Agerskov, "High-Strength Bolted Connections Subject to Prying," *Journal of the Structural Division, ASCE*, Vol. 102, ST1, January 1979.

Chapter Eighteen

Beam-To-Column Connections

18.1 INTRODUCTION

Beam-to-column connections play an important role in the load partition of structural frames. The major function of these connections is to transfer the loads that are applied to the beams and the floor system to the columns. In its simplest form, the connection is used to transfer only the end reaction of the beam to the column, and the beam is assumed to be simply supported. If restraints are provided, the end rotations of the beam are minimized, and the maximum positive moment in the beam can be reduced by the resulting end moments. Connections of this nature are often referred to as moment-resistant joints. Connections that are only capable of transferring the reaction of the beam are called shear connections.[18.1]

The behavior of beam-to-column connections is of major interest to engineers, and a significant amount of research has been done or is underway. In one category studies are aimed at developing and improving design rules for the beam-to-column connection.[16.1-16.3,18.1-18.22] This work focuses on the general requirements for connections, that is, (1) sufficient strength, (2) adequate rotation capacity, (3) sufficient stiffness, and (4) economical fabrication. The role of the beam-to-column connection in overall frame behavior is also of interest, and the prediction of the moment versus rotation characteristics of typical connections is a subject of recent and current study.[18.23-18.28]

Most of the early research on beam-to-column connections was performed on welded or riveted specimens. However, as the advantages of bolted connections and combination bolted and welded connections became more apparent because of decreased fabrication and erection costs, research on these types of connections was increased.

In current practice shop connections are often welded and field connections bolted. As a result of these fabrication procedures, a wide variety of beam-to-column connections are encountered in the field. It is still not possible to accurately describe and predict the behavior of many of these connections because of their complexity. This chapter summarizes the present state of knowledge and provides guidelines for design. The design recommendations for these joints are based on available information and result in a conservative, safe design. The ongoing experimental and theoretical work will permit the development of more liberal and improved design rules.

18.2 CLASSIFICATION OF BEAM-TO-COLUMN CONNECTIONS

Depending on their rotational characteristics, beam-to-column connections are classified as flexible, semi-rigid, or rigid connections.[18.1] Flexible connections are also called shear connections, and the semi-rigid and rigid-type connections are often referred to as moment-resistant connection.

The rotational characteristics of beam-to-column connections are important to the engineer because they affect the required beam size. For idealized rigid joints, the beam size is generally governed by the fixed end moment: for example, $M = wl^2/12$, for a uniformly loaded beam. If the same beam is attached to the column by a flexible-type shear connection, the maximum moment for the same loading case is $M = wl^2/8$. Actual situations in the field will generally be somewhat less rigid than assumed for the rigid connection and somewhat more rigid than assumed for the flexible connection. The classification of a connection depends entirely on the joint geometry and loading conditions. Generally, it is not possible to define how a joint should be classified unless test results and experience are available.

The simplest type of beam-to-column connection is the flexible connection that provides relatively low resistance against rotations. Hence, the connection mainly transfers shear to the column. Typical examples that fall into this category are the web angle connection (sometimes called the standard beam connection), web structural tee, and seat angle connections, shown in Fig. 18.1a. The structural T-connections, end-plate connections, and flange plate connections, shown in Fig. 8.1c, are typical examples of beam-to-column connections with high moment resistance. By combining web angles or a T-section with a beam seat and tension flange plate or angle, a semi-rigid connection results that has a greater moment resistance than the flexible connection. Unfortunately, the degree of restraint is difficult to evaluate unless test data are available.

Typical moment versus rotation characteristics for several types of beam-to-column connections are shown in Fig. 18.2. These relationships, combined with the beam line concept (introduced in Ref. 18.1), are often used to estimate the moment that will be developed by a particular connection, span, and beam size. The beam line defines the relationship between the end moment and end rotation of a beam. If a beam is uniformly loaded and subjected to restraining end moments M, the end slope ϕ is equal to

$$\phi = \frac{1}{24}\left(\frac{wl^3}{EI}\right) - \frac{Ml}{2EI}$$

This relationship is plotted in Fig. 18.2. The intersection of the beam line and moment versus rotation curves for the various connections indicates the moment resistance expected under these conditions. For example, the standard web angle connection (connection A in Fig. 18.2) develops about 20% of the fixed end mo-

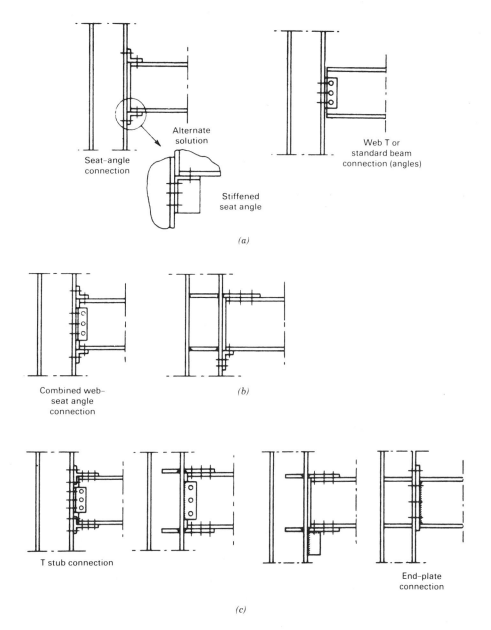

Fig. 18.1. Types of beam-to-column connections. Note. The need for column stiffeners in any of these connections must be checked. (*a*) Flexible connections. (*b*) Semi-rigid connections. (*c*) Rigid connections.

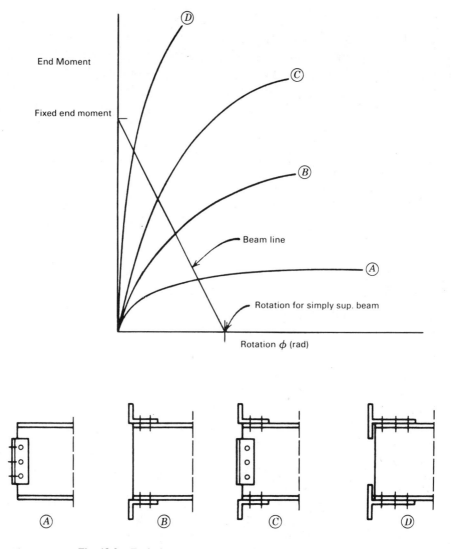

Fig. 18.2. Typical moment versus rotation curves and beam lines.

ment for this particular combination of beam and connection geometry. The same connection with added top and seat angles (connection *C*) develops about 75% of the fixed end moment.[18.1]

18.3   BEHAVIOR OF BEAM-TO-COLUMN CONNECTIONS

The stiffness and strength of beam-to-column connections are closely interrelated and of major importance to the performance of the connection. Strength requirements ensure that the connection has the ability to transfer the anticipated loads. Stiffness requirements relate to the ability to develop the desired restraint or lack of restraint. To meet the stiffness and strength requirements, additional stiffening of the column web or flanges may be needed, since certain joint components are subjected to highly localized, concentrated forces. Stiffeners are often necessary to prevent crippling of the column web in the compression region, excessive yielding of the column web, or deformation of the column flange near the tension flange of the beam. If the shear capacity of the column web is critical, shear stiffening may be required for that purpose as well.

The load versus deformation characteristics and approximate methods of analysis for typical beam-to-column connections are discussed in this section. Features from different types of connections are sometimes combined to meet the design requirements. Only the strength aspects of the connection are discussed in this section. Problems related to stiffening of the column web are treated separately (Section 18.4). The influence of the restraint characteristics on column or frame strength is not discussed in this Chapter.

18.3.1 Flexible Beam-to-Column Connections

The web angle or standard beam connection, as well as the seat angle connection, are typical flexible beam-to-column connections. Generally, they are assumed to be completely flexible and capable of transferring only shear. To justify these assumptions, the connections must allow for ample end rotation.

The rotation capacity of the connection is governed largely by the deformation capacity of the angles, as depicted in Fig. 18.3. Experiments have indicated that most of the rotation of the connection comes from the deformation of the angles; fastener deformations play only a minor role.[18.1,18.2] To minimize rotational resistance, the thickness of the angle should be kept to a minimum and a relatively large gage, g, provided (see Fig. 18.3).

A typical moment versus rotation diagram for a standard web connection that used both bolts and rivets is shown in Fig. 18.4. In this test, the heels of the angles on the tension side began to separate from the column flanges at about 260 kip-in. The toes of the angles remained in contact with the column. Yielding of the angles decreased the rotational resistance. After the compression flanges of the beams had made contact with the column flanges, the moment resistance of the connection increased, as shown in Fig. 18.4. Failure of the connection occurred from excessive yielding and tearing of the connection angles (see Fig. 18.5).

From this test series it was concluded that web angle beam-to-column connections offer some resistance to rotations at the ends of the beam. This partial restraint is relatively small and estimated to be about 10% of the fixed end moment provided by rigid moment-resistant connections.[18.2,18.28] Rotation restraints of the same order of magnitude can be expected in seat angle connections as well.[18.3] Jones *et al.* have provided a useful review (through 1980) of test data for various types and configurations of connections and show how a B-spline fit of data can be used to provide a good representation of the load versus deformation characteristics.[18.27]

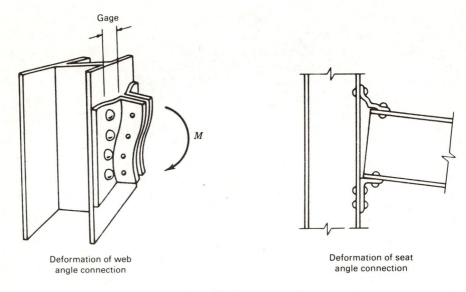

Fig. 18.3. Deformations of flexible beam-to-column connections.

Most web angle connections are checked only for their shear-carrying capacity, that is, the relatively small amount of moment present is neglected. This shear capacity can be governed by (1) the shear capacity of the fasteners, (2) the bearing capacity of the material adjacent to the bolts (angle legs adjacent to both column flange and beam web), including a check of end and edge distances, and (3) the shear capacity of the angles. Fasteners are assumed to be subjected to shear forces only; the tensile forces introduced by deformation of the angles (Fig. 18.3) are neglected. However, the effect of shear forces acting eccentrically should be included unless distances are small. The usual assumption is to consider the bolt group in the web as acted upon by an eccentric shear (Fig. 18.6), although work by Richard *et al.* on single plate framing connections indicates that this may not be a large enough allowance.[18.20]

The examination of end and edge distances for the fasteners should recognize that the rotation of the beam will result in the type of behavior shown in Fig. 18.7a. The upper bolts in the group will tend to push out material toward the end of the beam, and the lower bolts will tend to push out material toward the toes of the angles. It would be conservative to use the distance e_1 in checking the bearing capacity of the beam web and the lesser of e_2 and e_3 for the angles (see Fig. 18.7b).

The special case of coped beams should be recognized. Coping (or cutting back) of a flange might be necessary when a beam is to be connected to a column web or when a beam-to-girder connection requires that the top flanges be kept at the same elevation (Fig. 18.8a). In such a case, it is evident that a new mode of failure

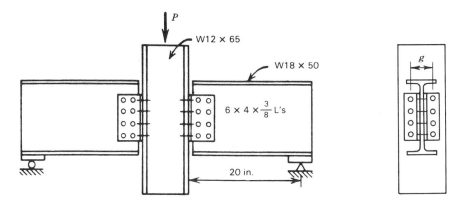

All fasteners $\frac{3}{4}$ in. dia.

Beam connection riveted,
column connection bolted.

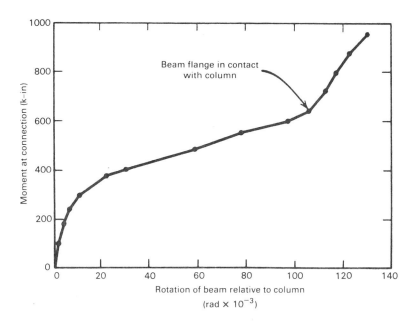

Fig. 18.4. Load versus deformation behavior of standard beam connection (Ref. 18.2).

Fig. 18.5. Angle failure in standard beam connection described in Fig. 18.4. (Courtesy of University of Illinois.)

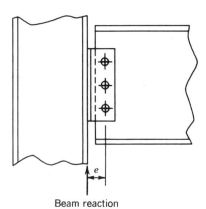

Beam reaction

Fig. 18.6. Eccentric shear acting on bolt group.

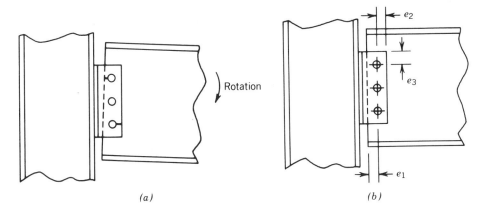

Fig. 18.7.   Effect of beam rotation on bolts. (*a*) Actual; (*b*) idealization.

is possible, the removal of a block of material, as indicated in Fig. 18.8*b*, by a combination of shear and tensile forces as the beam rotates relative to the angles. It has been found[18.17–18.19] that a good representation of the ultimate strength is given by the sum of tensile resistance on the horizontal surface a-b (Fig. 18.8*b*) and the shear resistance on the vertical surface b-c. Conservatively, the shear resistance could be used over the whole length a-b-c. Of course, the effect of the cope on the strength and stability of the beam also should be examined[18.38].

Instead of the double angles attached to the beam web that have been described thus far, a single angle or a single plate on one side of the beam web can be used. Obviously, there can be a saving in material (although the single element must be relatively thicker than either component of a double element), but more important savings usually result from the reduced cost of erection. It is much quicker to erect a beam that can be moved in laterally to a single connection piece than to bring a beam web into position between two connection pieces. This type of connection has received considerable attention recently.[18.13,18.20,18.21]

The framed beam connection has elements subjected to flexure (the outstanding legs of the web angles, especially) that give it ductility and that greatly contribute to fulfillment of the assumption of no (or, at least, little) restraint. A single plate framing connection has no comparable component. The ductility in this type of connection must come from the shear deformation of the bolts, from hole distortion (in the beam web or in the plate), and from out-of-plane bending of the plate. Of course, bolt slip prior to bearing might also be present, but this cannot be relied upon. Thus, although designers usually consider the single plate framing beam connection to be a flexible connection, care must be taken to ensure that sufficient ductility does exist so that the design assumption is satisfied.

For a single plate framing connection of the type shown in Fig. 18.9, the design requirements include selection of the bolts (shear capacity, checking the bearing

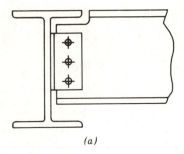

(a)

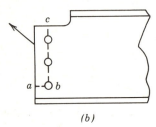

(b)

Fig. 18.8.   Effect of beam cope on failure. (a) Actual;
(b) tear-out.

stresses in the plates adjacent to the bolts, proper end and edge distances), pro-
portioning of the weld at the beam-to-beam or beam-to-column junction, and selection of a suitable framing plate. Usually, the latter requires only the selection
of the plate thickness; the other dimensions are controlled by the requirements for
weld length and by the number of bolts and their spacing. If an appreciable portion
of the end plate is unsupported, buckling of the plate should be investigated.

Richard and his coworkers conducted an extensive analytical and experimental

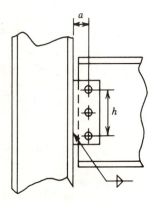

Fig. 18.9.   Single plate framing connection.

study of single plate framing connections, including an analysis of test results by others.[18.20,18.21] In order that ductility of the connection be ensured, they recommend that bolt shear or transverse tension tearing of the plate in line with the bolt (see Fig. 5.33a) not be permitted as failure modes. In other words, bearing deformations in the plate or in the beam web should be used as the principal mechanism to produce ductility.

Richard et al. recommend that the ratio L/d (see Fig. 5.35c) be at least 2.0. This is consistent with the recommendations given in Chapter 5 and will ensure that splitting-type failure will not occur. Next, in order to establish high bearing stresses and, therefore, relatively large bearing deformations before the maximum force is reached, it is recommended that the thickness of the thinner plate element (beam web or framing plate) not exceed about 50% of the bolt diameter when A325 bolts are used or about 70% of the bolt diameter when A490 bolts are used.

For the case of a uniformly loaded beam, and using the results of about 1500 beam line analyses of various beam sizes, bolt diameters, and bolt arrangements, Richard et al. established a ratio $(e/h)_{ref.}$ for beam span to depth ratios equal to or greater than 6 as follows:

$$(e/h)_{ref.} = 0.035 \text{ (beam span/beam depth)} \qquad (18.1)$$

A modified e/h ratio is then calculated as follows:

$$\frac{e}{h} = \left(\frac{e}{h}\right)_{ref.} \left(\frac{n}{N}\right) \left(\frac{S_{ref.}}{S}\right)^{0.4} \qquad (18.2)$$

where e is the connection eccentricity and h is the depth of the connection between extreme fasteners, that is, $h = (n - 1) \times p$, using

$$n = \text{number of bolts per vertical line}$$
$$p = \text{bolt pitch}$$

The other terms in Eq. 18.2 are

N = a numerical coefficient, to be taken as 5 for $\frac{3}{4}$-in. and $\frac{7}{8}$-in. dia. bolts and 7 for 1-in. dia. bolts
S = section modulus of the beam
$S_{ref.}$ = a numerical modifier, to be taken as 100 for $\frac{3}{4}$-in. dia. bolts, 175 for $\frac{7}{8}$-in. dia. bolts, and 450 for 1-in. dia. bolts.

(Modifications to accommodate cases of concentrated loadings are given in Ref. 18.20.)

The moment at the bolt line is given by

$$M = Ve \qquad (18.3)$$

and the moment at the weld line is given by

$$M = V(e + a) \tag{18.4}$$

where V is the shear force at the end of the beam and a is the distance from the weld line to the fastener line (see Fig. 18.9). The bolts, which had been selected by trial, can now be checked according to the procedure given in Chapter 13. The capacity of the weld and the plate itself can also be checked against the forces identified herein.

The procedure outlined above is believed to be satisfactory for single lines of bolts, either A325 or A490, using connected material with a yield strength of up to about 50 ksi. The use of more than a single line of bolts or the use of deep connections will be self-defeating. These arrangements will inevitably be stiffer than desirable for a flexible end connection. Design rules are also available for the case when A307 bolts are used.[18.21]

The upper angle in a seat connection (see Fig. 18.3) is mainly used to provide lateral stability for the beam. This component of the joint is not considered as load-carrying. The total shear force is assumed to be transmitted to the column by shear on the fasteners in the seat angle. The thickness of the seat angle is governed by critical bending stress on the outstanding leg. The usual practice is to consider the stress at the toe of the fillet of the outstanding leg. The required angle thickness is determined from the bending moment at that section. The reaction is assumed to act at the midpoint of the bearing length.[13.11]

18.3.2 Semi-Rigid Connections

There has been relatively little experimental work explicitly directed toward an understanding of the strength and deformation characteristics of semi-rigid connections. Most attention, particularly in the modern era, has been directed toward connections designed to be either flexible or rigid, with the recognition that neither of these ideals is exactly attainable. As was noted in Subsection 18.3.1, there has been a good deal of attention paid to the effect of all types of connections—flexible, semi-rigid, and rigid—upon the column strength.[18.23–18.27] The only type of semi-rigid connection that will be discussed in this section is the combined web-seat angle arrangement shown in Fig. 18.1b.

A combination web angle and seat angle connection results in significant increases in the joint restraint characteristics. Depending on the dimensions of the joint components and the loading conditions, these combination joints are sufficiently stiff to result in a substantial reduction in the midspan moment of a beam.[18.1]

Little experimental evidence is available on the load versus deformation behavior and load partition for this type of connection.[18.1] Since the behavior of the connections is complex and because of the lack of experimental data, a simplified, conservative approach is used for design. Current practice is to assume that the web angles will carry the shear. Thick top and bottom angles are used to transfer the end moment of the beam. Connections designed on the basis of these assumptions have provided satisfactory performance.

The design procedure for a shear connection is identical to that used for the web angle connection discussed in Subsection 18.3.1. The angles connecting the beam flanges to the column in the semi-rigid connection are considered to be load-carrying components; this was not the case for seat angle connections. Both angles are subjected to bending forces. However, the angle that connects the beam tension flange to the column flange is the critical one. A typical deformation condition for the tension angle is shown in Fig. 18.10*b*. Depending on the stiffness of the angle, prying forces may develop near the toe of the outstanding leg. Therefore, it is desirable to consider the influence of prying forces on the bending stress in the angle and the fastener tension. For analysis, the angle can be assumed to act like a T-stub connected to a rigid base and loaded in tension. This provides a conservative design because it assumes the angle to be fastened to a rigid base. Since the angle is fastened to a column flange, the decreased stiffness actually tends to relieve part of the restraint supplied by the angle. In general, the forces developed in a semi-rigid connection cannot be approximated in a reasonable way unless a test is conducted. This permits the stiffness and distribution of the forces in the connection to be evaluated.

The moment capacity of the connection is limited by the number of fasteners that can be placed in a single transverse line in the vertical leg of the angle connecting the tension flange to the column flanges. Because of deformation of the column flange (see Fig. 18.11), only the first fasteners on each side of the beam web may be fully effective in transferring the forces. Stiffening of the column flanges may be required unless they are at least as thick as the angle.

18.3.3 Rigid Connections

Replacing the angles of a combined web-seat angle connection (see Fig. 18.1*b*) with structural T-sections results in a connection with significantly increased mo-

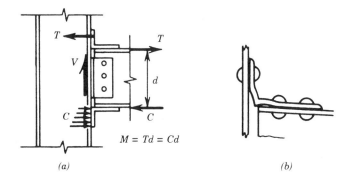

$$M = Td = Cd$$

(a)   *(b)*

Fig. 18.10. Assumed behavior of semi-rigid connection.

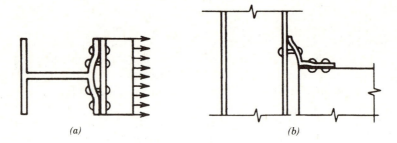

(a)   *(b)*

Fig. 18.11. Influence of deformations on fastener elongations.

ment resistance. Such a connection (see Fig. 18.1*c*) provides a rigid joint with high rotational resistance. The increase in rotational resistance is provided by the symmetrically loaded T-sections. Unlike angle connections, which are connected to the column flanges by two or more fasteners on one line, the T-section allows two or more fasteners to be used effectively on two lines to transfer the tensile forces that result from the applied moment. This results in an increase in moment capacity and joint stiffness. Since the T-sections are symmetrically loaded, they do not permit as much deformation to occur as compared with eccentrically loaded angles (see Fig. 18.3).

The design of the T-stub connection utilizes assumptions similar to those used for combined web-seat angle connections. The flange connection is assumed to transfer the moment, and the shear force is transferred by the web connection. Tests were carried out on connections of this type to evaluate the validity of these assumptions,[16.1,16.2] and typical test results are illustrated in Fig. 18.12. The effect of beam shear and the presence of the web angles on the behavior of the flange connections was investigated. In addition, these tests yielded valuable information on the rotation capacity of these connections.

The test results indicated that the behavior of the bolts connecting the T-stubs to the beam flanges was similar to the behavior observed in simulated flange plate splice tests.[16.2] The connection strength exceeded the plastic moment of the gross cross-sectional area of the beam, despite the presence of the holes in the flanges. Substantial rotational capacity was attained (see Fig. 18.12) when premature failure of the joint components was prevented. It was further concluded that the beam shear had no significant effect on the performance of the connection. The shear was largely carried by friction between the T-stubs and the column flanges. There was very little difference in bolt tension in the individual bolts connecting the tension T-stub, regardless of the magnitude of the prying forces.[16.2]

The test results generally supported the assumptions made in design. Although some shear can be transferred by the web of the T-stub, web angles are needed to assist with the shear transfer. This is particularly true if large shear forces exist.

In current (1987) steel fabrication practice, it is probably more common to use

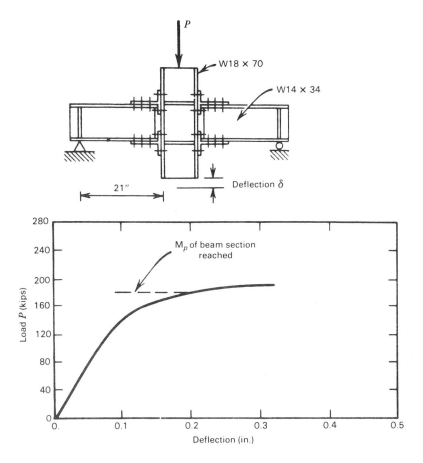

Fig. 18.12. Load versus deflection curve for a T-stub connection (Ref. 16.1).

a flat plate, groove-welded to the column flange, in place of the T-stub. This produces a simpler and more compact connection. The groove weld can be made in the shop and the bolts between the plate and the flange of the beam installed in the field. Web framing angles on one or both sides of the web or shear plates can be used to transfer the shear between the beam and the column. A single plate welded framing connection on one side of the web is the most common method used to transfer the shear between the beam and the column. The flange connections prevent the large rotations experienced by the single plate connections in simple beams.

Chen and his co-workers conducted a series of tests on various types of bolted beam-to-column moment connections[18.8, 18.9] and compared them to fully welded connections.[18.29] (In all cases, the flange plates were groove-welded to the columns.) The results that will be discussed here are only those in which the connection was made to the column flanges; tests were also conducted on beam-to-column web connections.[18.10]

Specimens were designed using the assumption that the flanges carried all of the moment and the web carried all of the shear. (One test was carried out on a connection that had no connection between the beam web and the column flange, that is, the groove welds at the beam flange level were expected to transfer both shear and moment. This connection exhibited neither adequate strength nor ductility.) The bolted parts of two of the specimens tested were designed as bearing-type connections, and a third specimen used a slip-resistant connection. A fourth test used a stiffened beam seat in addition to the flange and web details described herein; it will not be discussed.

Figure 18.13 shows the behavior of these connections.[18.8] In Fig. 18.13a a "fully bolted" connection, C7, is compared with a fully welded, but otherwise comparable, connection. Fully bolted means that the web shear plate and the flange connection plates were bolted to the beam. For this specimen, the bolts were designed as bearing-type. A490 bolts of 1-in. dia. installed in $1\frac{1}{16}$-in. dia. holes were used in both the flange and the web connections. Two responses representing theoretical cases are shown: one includes strain-hardening and the other does not. The fully welded connection follows the theoretical prediction that includes strain-hardening quite closely, except that there is a rounded knee, as would be expected, due to yielding. The response curve of the bolted connection shows a change in slope at about 150 kips, probably due to slip of fasteners as well as yielding. Both the ultimate strength and the rotational capacity of the bolted connection were greater than that of the fully welded connection.

In Fig. 18.13b the behavior of a fully welded connection and two otherwise comparable bolted connections are compared. One of the bolted connections used a bearing-type design (C9), and the other used a slip-resistant design (C8). In Specimen C9, A490 bolts of 1-in. dia. were installed in $1\frac{1}{16}$-in. dia. holes to connect the flange plate to the beam flange. The web connection was made using $\frac{3}{4}$-in. dia. A325 bolts. Slotted holes $1\frac{7}{8}$ in. long were used horizontally in the single web shear plate. A covering bar was used, in accordance with the RCSC specification. The fastener and hole arrangement for specimen C8 was similar, except that $1\frac{1}{4}$-in. dia. holes were used for the 1-in. dia. A490 bolts.

Figure 18.13b shows that the response curve that uses strain-hardening once again provides a good representation of the actual response of the fully-welded connection. As with specimen C7, this bearing-type bolted connection (C9) also shows two distinct slopes in the initial region. The slip-resistant connection (C8) followed the theoretical curve more closely and did not show any distinct change in slope.

The slip that occurred in the tests of these bolted connections was a series of small individual slips that took place in the second-slope region of the load versus deflection curves. Shown in each of Fig. 18.13a and b is a horizontal line illustrating the effect of the horizontal slips upon the vertical deflection. This horizontal

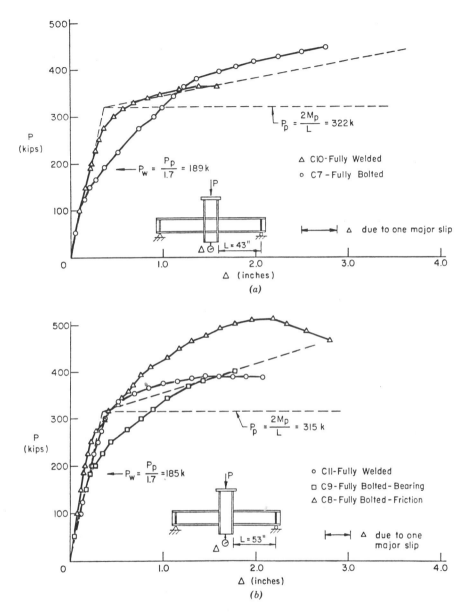

Fig. 18.13.   Behavior of fully welded and fully bolted rigid connections (Ref. 18.8).

line represents the effect that would have occurred had there been one major slip, rather than the many minor slips that actually took place.

Whether the load at which the second-slope of the load versus deflection response curve will be above or below the working load of the connection depends on the particular details involved. Standig *et al.*[18.8] have shown that this load level can be predicted with reasonable accuracy using the principles outlined in this *Guide*. The connection of the beam flange to the moment plate is idealized as a slip-resistant lap joint, and the slip load obtained by this analysis then can be compared with the theoretical force in the flange plate (at the first line of bolts from the free end), assuming that all moment is carried by the beam flanges.

All three bolts specimens tested exhibited adequate rotational capacity as compared with the fully welded joints.

End plates welded to the beam cross-section have been used in beam-to-column connections and butt-type beam splices (see Chapter 16). Two types of end plates are used, as shown in Fig. 18.14. In one type the fasteners are placed only between the beam flanges, and in the other type the end plate is extended beyond the tension flange and fasteners are centered around the flange. Sometimes, this flange extension is stiffened.[18.22]

The exact load transfer in this type of connection is complex. The shear forces acting on the connection are transferred by frictional resistance and/or by shear on the fasteners. The fasteners are also subjected to tensile loads that resist the bending moment. The forces in the bolts change under the applied loads and are dependent on the magnitude of the initial bolt tension.

The end-plate connection is an economical way of fastening beams to column

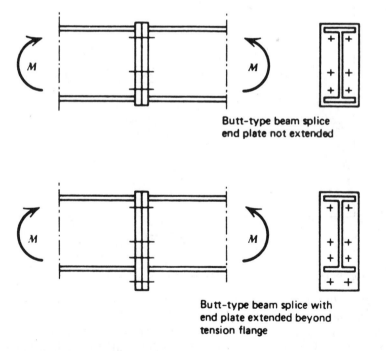

Butt-type beam splice
end plate not extended

Butt-type beam splice with
end plate extended beyond
tension flange

Fig. 18.14. End-plate types. *Note*. Connect end plates to beams with enough weld to develop full bending strength of beam.

flanges, and a number of studies, both analytical and experimental, have been carried out in recent years.[16.1, 16.2, 18.4–18.6, 18.11, 18.14–18.16] The problem is a complex one because the end-plate connection is highly indeterminate. The most recent analytical studies have used the finite element method to study the distribution of internal forces. Because of the relatively confined physical system involved, experimental studies have generally involved only the measurement of the moment versus rotation response of the connection and, in some cases, the measurement of bolt forces.

The end-plate connection has a number of similarities to the T-stub connection just discussed. It is evident on the basis of that examination that the following potential critical regions or effects will have to be examined for an end plate connection:

1. Buckling, crippling, or yielding of the column web opposite the beam flange that delivers the compressive force
2. Yielding of the column flange (or excessive deformation of the column flange) opposite the beam flange that delivers the tensile force
3. Yielding or fracture of the connectors (welds or bolts)
4. Failure of the end plate itself due to yielding or fracture
5. Yielding due to shear in the panel zone of the column web

This list is intended to cover the situation wherein a beam or girder is framed into the flange of a column. The situation will be somewhat different for a beam or girder splice that uses an end-plate connection. End-plate connections between beams or girders and a column web are not generally used.

A number of experimental studies have been made to examine the load versus deformation behavior of this type of connection and to develop design rules.[16.1, 16.2, 18.4–18.6, 18.11] These studies have indicated that the bolts mainly effective in resisting the tension flange force are those adjacent to the tension flange. This is illustrated in Fig. 18.15 where the bolt forces in a moment splice end plate connection are plotted as a function of the applied load.[16.2] The measured bolt forces were all similar at the start of the test. As load was applied, the forces in the bolts centered about the tension flange (levels 3 and 4) increased from about 30 kips to about 48 kips. The forces in the bolts at level 2, close to the neutral axis of the beam, showed no appreciable change as the load was applied, and the bolts on the compression side, level 1, showed a decrease in force from about 28 kips to 16 kips.

It was concluded from Fig. 18.15 that the variation of the force in the several rows of a bolt pattern depends primarily on the stiffness of the end plate and whether the plate yields before fracture of the critical fasteners takes place. At first, strains will increase in proportion to the distance of the fasteners from the compression flange. Because of the strain gradient, differences in bolt loads result, but these differences will decrease as plastic deformations of the bolt develop. If the bolts have sufficient ductility, all bolts in the tension region will develop the same

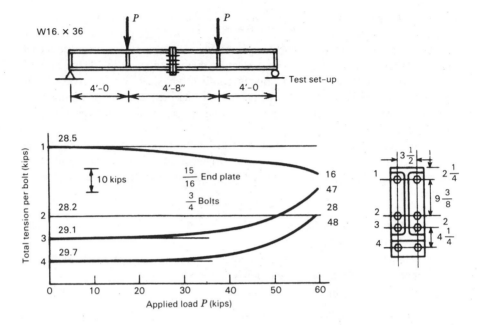

Fig. 18.15. Bolt force versus applied load (Ref. 16.2).

capacity at ultimate load.[16.2] Unless it is sufficiently thick, the end plate will yield and a linear strain distribution will not occur. This is apparent in Fig. 18.16, which shows an end-plate connection after failure.[18.5] The pressure distribution at the interface of the end plate and the column is shown in Fig. 18.17 and indicates that prying forces were developed at the edges of the end plate near the tension flange.[18.5]

Test results have shown that the bolts that are effective in resisting the moment for flexible end-plate connections are those adjacent to the tension flange. The connection is flexible if prying forces are developed at the edge of the end plate in the tension region. If a connection is designed such that no prying forces are developed, a linear strain distribution among the fastener rows can be assumed, and the inner fasteners may contribute to the capacity of the connection. The ultimate moment resistance of the connection is the summation of the products of the effective fastener loads and their respective distance from the center of rotation. At the ultimate load the center of the rotation is near the centerline of the compression flange. This is compatible with existing experimental observations.[16.2, 18.4, 18.5]

The design of end-plate connections requires that the connection provide adequate strength, that is, both the size and number of bolts and the end-plate thickness must be satisfactory, and that there be adequate rotation capacity such that the desired moment capacity can be attained. In addition, the connection must be stiff enough so that permanent deformations are not introduced under working loads.

Fig. 18.16. End-plate connection after failure. (Courtesy of University of Sheffield.)

On the beam side of the end-plate connection, the weld between the end plate and the beam must be proportioned. On the column side of the connection, the delivery of the shear and moment from the beam must be accomplished with strength and stability requirements being satisfied.

Krishnamurthy has reported the results of an extensive analytical study of end-plate connection behavior.[18.11] Two- and three-dimensional finite element analyses of many T-hanger and end-plate connections were carried out. Krishnamurthy noted that there can be significant differences in stiffness between the assembly at the bolt line and at the face of the end plate-to-beam flange junction (the "load line"), and he also observed that there is a significant difference in stiffness in the end plate in the region where it is extended beyond the beam flange and in the region of the end-plate between the beam flanges. As a consequence of these and other

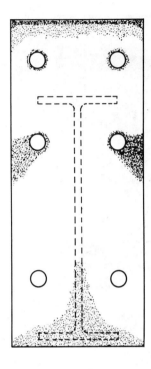

Fig. 18.17.   Pressure distribution at interface as recorded on interposed paper backed up by carbon paper (Ref. 18.5).

factors, Krishnamurthy proposed a method of analysis based on a traditional approach but modified on the basis of his analytical studies to account for these effects.

The analysis of an end-plate connection, a highly redundant system, must be subjected to physical testing as well as to analytical testing. In Ref. 18.11 Krishnamurthy reported the results of 10 tests of end-plate connections (9 are reported in the main body of the report, and 1 in an appendix). In one test, the moment attained was well below the plastic moment capacity of the section (58%) because torsional buckling occurred. In two others, the presence of a slender web meant that local buckling occurred before the moment capacity of the section could be reached. Thus, it can be observed that there were seven tests that could be used to substantiate the method proposed by Krishnamurthy. For these tests, Krishnamurthy reported that the ratio p_e/d ("effective distance" from the bolt line to the load line/the bolt diameter) ranged from 0.8 to 1.4. Dismissing the results of certain tests as invalid as a suitable measure of the end-plate behavior (see above), this ratio only extends over the range of 0.8–1.1. Indeed, the mean value of the ratio is 1.0, with a standard deviation of 0.1. Furthermore, the bolt diameters in these tests were about two times the thickness of the end plate. Thus, the situation is one in which the physical tests represent the case of relatively large bolts connecting a relatively thin end plate, and wherein the bolt is located very close to the beam flange-to-end plate junction. (In most cases, the Krishnamurthy test specimens used a bolt arrangement that was at or slightly above the limit considered to be a practical minimum.[18.30]) It could be expected that these tests would not result in any bolt failures, and none were reported.

The procedure for the design of end plates and their fasteners as recommended by Krishnamurthy[18.11] and as adopted in the eighth edition of the AISC *Steel Construction Manual*[18.31] is as follows:

1. Assume that the beam flanges carry all of the moment, and calculate the force in each flange accordingly.
2. Determine the size and number of bolts required to transfer the flange force. No allowance is made for additional forces (above the nominal values) due to prying action.
3. Calculate an effective distance (p_e) between the line of action of the bolt line and the "load line" at the end plate-to-beam flange junction (see Fig. 18.18) as

$$p_e = p_f - 0.25d_b - w_t \qquad (18.5)$$

4. Calculate the moment in the end plate (M_t) at the location of the load line, using the bolt forces and the effective distance calculated above. The load line location is assumed to be a location of fixed end moment.
5. Compute a modified moment at this location as

$$M_d = \alpha_m M_t$$

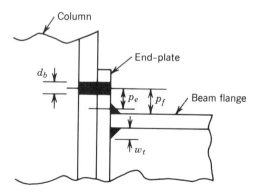

Fig. 18.18. Nomenclature for end plate design.

where

$$\alpha_m = 1.29 \left(\frac{F_y}{F_{bu}}\right)^{0.4} \left(\frac{F_{bt}}{F_p}\right)^{0.5} \left(\frac{b_f}{b_s}\right)^{0.5} \left(\frac{A_f}{A_w}\right)^{0.32} \left(\frac{p_e}{d_b}\right)^{0.25} \qquad (18.6)$$

and

F_y = yield strength of beam and plate material
F_{bu} = ultimate tensile strength of bolt material
F_{bt} = allowable tensile strength of bolt
F_b = allowable bending stress in end plate
b_f = width of beam flange
b_s = width of end plate
A_f = area of tension flange of beam
A_w = area of web of beam
p_e = effective bolt distance (see Eq. 18.5)
p_f = distance of bolt from face of beam flange
d_b = bolt diameter
w_t = throat size of fillet weld
w_s = size of fillet weld.

6. Calculate the end plate thickness (t_s) using simple bending theory.
7. Compute an effective maximum end plate width as

$$b_e = b_f + 2w_s + t_s \qquad (18.7)$$

If b_e is less than b_s, recalculate t_s using the value of b_e in place of b_s.
8. Check the shear stress in the plate.

Design aids are available that simplify the calculations required.[18.31] The method proposed by Krishnamurthy and adopted by the AISC will result in thinner end plates than designers have been accustomed to in the past. Agerskov[18.32] has commented that bolt prying forces are likely to be present in end-plate connections and should not be ignored, and that, with the thinner end plates, deformations between yield moment and the plastic moment levels might become excessive. Similarly, McGuire[18.33] suggests that there might be a degradation of bolt clamping forces even under working load levels because of the thin end-plates. Mann and Morris, who have analyzed end-plate connections using a yield line approach, recommend that an increase of $33\frac{1}{3}\%$ over normal bolt load levels be applied in recognition of bolt prying forces.[18.14] In the face of these criticisms, it must be noted that light end-plate connections have been used successfully in the industrialized metal building industry for many years.[18.34]

The design procedure developed by Krishnamurthy appears to give satisfactory results within the parameters examined, especially the use of bolts that have a diameter that is large relative to the end-plate thickness and that are placed (effective distance p_e) no further than about one bolt diameter from the load line at the end plate-to-beam flange junction. When many repetitions of a connection type are required, this procedure will be advantageous because it reduces material cost. On the other hand, when the number of connections is not large, a reduction in end-plate thickness may not be significant since the labor component will not be much reduced over that for a thicker end plate. In these cases, the designer and fabricator might prefer a more conservative approach since it provides more leeway in detailing the connection and, thereby, in ease of fabrication and erection. The bolts and end plate adjacent to the tension flange can be conservatively designed by assuming that they are equivalent to a T-stub connection loaded in tension. Design procedures for this idealization are given in Chapter 17.

Although the primary transfer of shear is concentrated near the compression side of the joint, it can be conservatively assumed that all bolts carry an equal part of the shear load. Hence, the fasteners in an end-plate connection are subjected to combined shear and tension. The magnitude of initial clamping force does not influence the ultimate strength of the connection; it does influence the shear resistance of slip-resistant joints.

When end plates do not extend beyond the tension flange, their behavior is not well known because available data are not extensive. In general, these types of end-plate connections are less efficient and require thicker end plates. Reference 16.2 suggested that end plates that do not extend beyond the tension flange should be proportioned to resist a moment equal to the product of the beam flange force and the distance between the center of the beam flange and the nearest row of bolts. Plate thicknesses determined in this manner appear to provide a linear variation in fastener strain throughout the connection depth. Additional test data are needed to verify this suggested method for a range of sizes.

All of the foregoing discussion on end-plate connections has assumed that the bolts adjacent to the tension flange of the beam will be arranged in two lines, one just above the beam flange and one just below. If this arrangement does not provide a sufficient number of fasteners, it may be necessary to use more than one bolt line. Extension of the end plate above the tension flange of the beam to accommodate two bolt lines is practical only if a very thick end plate is used or if the end-plate extension is stiffened. Work on stiffened end plates has been reported by Murray and Kukreti.[18.22]

18.4 STIFFENER REQUIREMENTS FOR BOLTED BEAM-TO-COLUMN CONNECTIONS

The full capacity of a moment-resisting beam-to-column connection can only be developed if the column does not exhibit premature failure. The column is subjected to highly localized forces resulting from the applied moments and can deform as shown schematically in Fig. 18.19a. Excessive deformations of connected parts should be avoided. There are two major effects of the beam flange forces that have

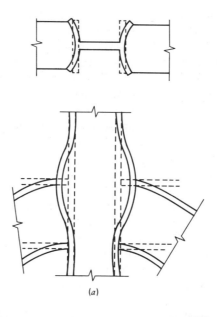

Fig. 18.19. Deformation of column in moment resistant connection. (*a*) Distortion of unstiffened column. (*b*) Web crippling in beam-to-column connection. (Courtesy of British Steel Corp.)

to be examined because they may result in excessive deformations. On the compression side of the beam, crippling or overall buckling of the column web can occur. On the tension side, excessive yielding and distortion may result in fracture of the column web or bolts. Web buckling is illustrated in Fig. 18.19b where an end-plate connection at ultimate load is shown. Because of the lack of stiffening in the compression region, the column web buckled and the connection could not develop the plastic moment capacity of the beam.[18.4]

Several investigators[18.4–18.6] have examined the stiffening requirements for bolted beam-to-column connections. Many joint geometries and boundary conditions exist; the problem is therefore extremely complex and no satisfactory general design approach is possible. Often the requirements developed for stiffening welded beam-to-column connections are used.[18.35–18.37] Since the concentrated forces are more localized in welded connections, application of the rules developed for welded connections to bolted connections results in a conservative design for the same moment capacity.

Standig *et al.*[18.8] and Huang *et al.*[18.9] have confirmed the adequacy of this approach for stiffening bolted connections. It is noted, however, that a bolted moment connection can have an actual moment capacity that is considerably larger than an all-welded connection designed for the same conditions. If advantage is to be taken of this increased capacity, stiffening requirements might require modification. Pending further research, criteria based in part on the requirements used for welded beam-to-column connections are reasonable.

The requirements for stiffening of the column are summarized as follows. As proposed in Ref. 18.35, the compression flange force on the column is assumed to be distributed on a 2.5:1 slope from the point of contact to the column k-line (see Fig. 18.20). If the compression flange force is distributed to the column flange by either an end plate or a structural T-section, it can be assumed to be distributed over a region on the column face about twice as great as the beam flange thickness. Hence, the force in the beam flange is assumed to be resisted by a length of column web equal to $(Q + 5k_c)$, where Q is the sum of the beam flange thickness and twice the end-plate thickness (for the plate connection) or the web thickness of the T-stub and twice its flange thickness, and k_c is the column fillet depth. For equilibrium, the resistance of the effective area of the web must equal or exceed the applied concentrated force of the beam tension or compression flange. This yields the following condition;

$$\sigma_{yc} w_c (Q + 5k_c) \geq A_f \sigma_{yb} \qquad (18.8)$$

where w_c is the thickness of the column web, and A_f is the flange area of the beam. The yield point of the column web is given by σ_{yc} and the yield point of the beam flange by σ_{yb}. If the column web resistance is less than provided by Eq. 18.8, stiffeners are required.

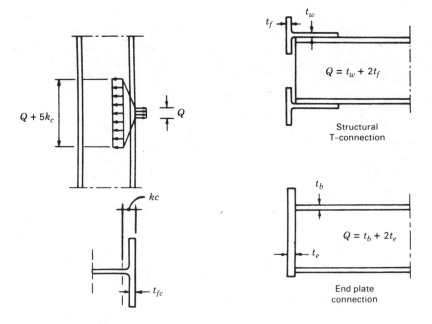

Fig. 18.20. Assumed distribution of compression flange force in bolted beam-to-column connections.

If flange splice plates are welded to the column on the compression or tension side of the beam, the provisions developed for welded connections are directly applicable.[18.35] The force from the compression flange is resisted by a length of the column web equal to $(t_s + 5k_c)$, where t_s is the splice plate thickness.

An upper limit must be placed on this analysis of strength because, at some value of the column web slenderness, the region in compression will buckle. Chen and Oppenheim have suggested that a square panel $(d_c \times d_c)$ can be used for the analysis.[18.36] Using plate buckling theory and the usual values for E and ν for steel, this upper limit can be expressed as

$$\frac{d_c}{w_c} \leq \frac{180}{\sqrt{\sigma_{yc}}} \qquad (18.9)$$

If the actual web slenderness of the column exceeds the value given by Eq. 18.9, a capacity prediction based on test data is given by[18.37]

$$P = \frac{4100\, w_c^3 \sqrt{\sigma_{yc}}}{d_c} \qquad (18.10)$$

For the tension flange, Ref. 18.35 has shown that the column flange provides adequate resistance against excessive deformations from the concentrated forces delivered by the tension splice plate if

$$t_{fc} \geq 0.4 \left(A_f \frac{\sigma_{yb}}{\sigma_{yc}} \right)^{1/2} \tag{18.11}$$

where t_{fc} is the column flange thickness. Tests of welded connections proportioned to these recommendations indicated that the connections were able to develop the full plastic moment of the beam.[18.9,18.35]

If a T-section or an end plate is bolted to the column flange, the concentrated tension force is distributed into the column flange by the fasteners. The system of applied forces differs significantly from the case of the splice plate welded to the column. The application of Eq. 18.11 is likely to yield overly conservative results.

European practice is to use a yield line analysis to examine the requirements for column flange thickness when bolts are used to deliver the load from the tension flange of the beam.[18.16]

When column stiffeners are required, they should be proportioned to carry the excess between the beam flange force and the calculated resisting capacity of the column web or flange.

If a single beam frames into a column or if the moments from two beams at an interior connection differ by a large amount, the web of the column can be subjected to large shears. In such situations it may be necessary to provide shear stiffening in the form of diagonal stiffeners or doubler plates. Design of such stiffeners is treated in many design handbooks.[13.11,15.1,17.6]

18.5 DESIGN RECOMMENDATIONS

Depending on the anticipated behavior, bolted beam-to-column connections are designed either as slip-resistant or bearing-type joints. The design recommendations in Chapter 5 for fasteners in butt joints are also applicable to the design of bolted beam-to-column connections.

The bolts in an end-plate connection are subjected to combined tension and shear. The elliptical interaction curve for bolts subjected to combined loading conditions (see Eq. 4.8) can be used to examine the adequacy of the fasteners.

With the exception of end-plate connections, it can be assumed for design that the web connection or the seat angle transfers the shear component. Web shear connections should be designed as eccentrically loaded joints in accordance with the recommendations given in Chapter 13. The moment on a beam-to-column connection is transferred by structural components connected to the beam and column flanges. The recommendations given in Chapter 16 for beam and girder splices are applicable to the design of the beam flange connection. The tension connection between the beam flange and column flange is usually critical for design.

Prying forces should be considered for the design of the fasteners as well as joint components. The bolts and end plate adjacent to the tension flange can be treated as an equivalent tee stub connection, loaded in tension. Design recommendations for the T-stub connection are given in Chapter 17. Alternatively, the method recommended by Krishnamurthy for the design of end plates and their connectors can be used.

Special attention should be given to the bending stiffness of the column flanges to which the T-section or the end plate is fastened. The deformations of the column flanges and the T-section (end plate) may introduce prying forces (see Chapter 17), depending on their stiffnesses.

Stiffening the column may be required to prevent premature failure of a joint component due to column web crippling or column flange deformation. For connections with flange splice plates welded to the column, the requirements for welded connections can be applied.[2.11,18.35] If the compression flange force is transferred through an end plate or a T-section, Eq. 18.8 can be used to determine whether additional column stiffening is needed.

$$\sigma_{yc} w_c (Q + 5k_c) \geq A_f \sigma_{yb} \tag{18.8}$$

For slender webs the stability of the compression region may govern rather than strength alone. Reference 18.37 has suggested that the following relationship (see Eq. 18.10) be satisfied when $d_c/w_c > 180\sqrt{\sigma_{yc}}$

$$w_c^3 \leq \frac{\sigma_{yb}}{\sigma_{yc}} A_f \frac{d_c\sqrt{\sigma_{yc}}}{4100} \tag{18.12}$$

where d_c is the column web depth.

The flanges of the column must not deform excessively under the action of the concentrated flange tensile forces. If splice plates welded to the column are used, adequate flange resistance is provided by

$$t_{fc} \geq 0.4 \left(A_f \frac{\sigma_{yb}}{\sigma_{yc}} \right)^{1/2} \tag{18.11}$$

For bolted T-connections in tension (including end-plate connections), the use of Eq. 18.11 will be conservative. A yield line analysis can be used as an alternative.

When stiffeners are required, they must be proportioned to carry the difference between the concentrated force calculated to be in the beam flange and the calculated resistance of the column web. For stiffeners opposite the beam compression flange, the required stiffener area can be determined from equilibrium. This yields

$$\sigma_{ys} A_{st} = \sigma_{yb} A_f - w_c (Q + 5k_c) \sigma_{yc} \tag{18.13}$$

If $C_1 = \sigma_{yb}/\sigma_{yc}$ and $C_2 = \sigma_{yc}/\sigma_{ys}$, Eq. 18.13 can be expressed as:

$$A_{st} = [C_1 A_f - w_c(Q + 5k_c)] C_2 \tag{18.13a}$$

If Eq. 18.12 governs the column web thickness, the required stiffener area becomes

$$A_{st} = \left[C_1 A_f - \frac{4100\, w_c^3}{d_c \sqrt{\sigma_{yc}}} \right] \sigma_{yc} \tag{18.14}$$

A comparable requirement can be developed for stiffeners opposite the tension flange by considering the needed additional flange area to be resisted by stiffeners. Equilibrium yields

$$\sigma_{ys} A_{st} = \sigma_{yb} A_f - \sigma_{yb} A_f' \tag{18.15}$$

where $A_f \sigma_{yb}$ is the actual beam flange tension force, and $A_f' \sigma_{yb}$ is the beam flange tension force that would not require stiffeners. This latter force can be estimated from Eq. 18.11 for the column flange thickness furnished. This yields

$$A_f' = \frac{\sigma_{yc}}{\sigma_{yb}} t_{fc}^2 \frac{100}{16} \cong 6 \frac{\sigma_{yc}}{\sigma_{yb}} t_{fc}^2$$

Substitution into Eq. 18.15 yields

$$\sigma_{ys} A_{st} = \sigma_{yb} A_f - 6\sigma_{yc} t_{fc}^2 \tag{18.15a}$$

Hence, the required stiffener area opposite the team tension flange becomes

$$A_{st} = [C_1 A_f - 6 t_{fc}^2] C_2 \tag{18.15b}$$

As a practical requirement, if stiffeners are required opposite both the beam tension and compression flanges, they are generally made the same size.

The fastener shear stresses and the bearing stresses suggested in Chapter 5 were shown in Refs. 18.7, 18.9, and 18.36 to be fully applicable to beam-to-column connections.

REFERENCES

18.1 J. C. Rathbun, "Elastic Properties of Riveted Connections," *Transactions, ASCE,* Vol. 101, 1936.

18.2 W. H. Munse, W. G. Bell and E. Chesson, Jr., "Behavior of Riveted and Bolted

Beam-to-Column Connections," *Journal of the Structural Division, ASCE,* Vol. 85, ST3 March 1959.

18.3 R. A. Hechtman and B. G. Johnston, *Riveted Semi-Rigid Beam-to-Column Building Connections,* AISC, Progress Report 1, 1947.

18.4 J. R. Bailey, "Strength and Rigidity of Bolted Beam-to-Column Connections," *Proceedings, Conference on Joints in Structures,* University of Sheffield, Sheffield, July 1970.

18.5 J. O. Surtees and A. P. Mann, "End Plate Connections in Plastically Designed Structures," *Proceedings, Conference on Joints in Structures,* University of Sheffield, Sheffield, July 1970.

18.6 A. N. Sherbourne, "Bolted Beam-to-Column Connections," *The Structural Engineer,* London, 1961.

18.7 J. S. Huang and W. F. Chen, *Steel Beam-to-Column Moment Connections,* Meeting Reprint 1920, ASCE National Structural Engineering Meeting, April 1973.

18.8 K. F. Standig, G. P. Rentschler, and W. F. Chen, *Tests of Bolted Beam-to-Column Moment Connections,* Fritz Engineering Laboratory Report No. 333.31, Lehigh University, Bethlehem, Pennsylvania, May 1975.

18.9 J. S. Huang, W. F. Chen, and L. S. Beedle, *Behavior and Design of Steel Beam-to-Column Moment Connections,* Welding Research Council Bulletin 188, New York, October 1973.

18.10 G. P. Rentschler, W. F. Chen, and G. C. Driscoll, "Beam-to-Column Web Connection Details," *Journal of the Structural Division,* ASCE, Vol. 108, ST2, February 1982.

18.11 N. Krishnamurthy, "A Fresh Look at Bolted End-Plate Behavior and Design," *Engineering Journal,* AISC, Vol. 15, No. 2, Second Quarter, 1978.

18.12 W. F. Chen and K. V. Patel, "Static Behavior of Beam-to-Column Moment Connections," *Journal of the Structural Division,* ASCE, Vol. 107, ST9, September 1981.

18.13 S. L. Lipson, "Single-Angle Welded-Bolted Connections," *Journal of the Structural Division,* ASCE, Vol. 103, ST3, March 1977.

18.14 A. P. Mann and L. J. Morris, "Limit Design of Extended End-Plate Connections," *Journal of the Structural Division,* ASCE, Vol. 105, ST3, March 1979.

18.15 P. Grundy, I. R. Thomas, and I. D. Bennetts, "Beam-to-Column Moment Connections," *Journal of the Structural Division,* ASCE, Vol. 106, ST1, January 1980.

18.16 J. Witteveen, J. W. B. Stark, F. S. K. Bijlaard, and P. Zoetemeijer, "Welded and Bolted Beam-to-Column Connections," *Journal of the Structural Division,* ASCE, Vol. 108, ST2, February 1982.

18.17 J. A. Yura, P. C. Birkemoe, and J. M. Ricles, "Web Shear Connections: An Experimental Study," *Journal of the Structural Division,* ASCE, Vol. 108, ST2, February 1982.

18.18 J. M. Ricles and J. A. Yura, "Strength of Double-Row Bolted-Web Connections," *Journal of Structural Engineering,* Vol. 109, No. 1, January 1983.

18.19 P. C. Birkemoe and M. I. Gilmor, "Behavior of Bearing Critical Double-Angle Beam Connections," *Engineering Journal,* AISC, Vol. 15, No. 4, Fourth Quarter 1978.

18.20 R. M. Richard, P. E. Gillett, J. D. Kriegh, and B. A. Lewis, "The Analysis and Design of Single Plate Framing Connections," *Engineering Journal,* AISC, Vol. 17, No. 2, Second Quarter 1980.

18.21 R. M. Richard, J. D. Kreigh, and D. E. Hormby, "Design of Single Plate Framing Connections with A307 Bolts," *Engineering Journal,* AISC, Vol. 19, No. 4, Fourth Quarter 1982.

18.22 T. M. Murray and A. R. Kukreti, "Design of 8-Bolt Stiffened End-Plate Moment Connections," Third Conference on Steel Developments, Australian Institute of Steel Construction, Melbourne, May 1985.

18.23 G. C. Driscoll, Jr., "Effective Length of Columns with Semi-Rigid Connections," *Engineering Journal,* AISC, Vol. 13, No. 4, Fourth Quarter 1976.

18.24 D. J. L. Kennedy, "Moment-Rotation Characteristics of Shear Connections," *Engineering Journal,* AISC, Vol. 6, No. 4, October 1969.

18.25 N. Krishnamurthy, H. T. Huang, P. K. Jeffrey, and L. K. Avery, "Analytical M-Theta Curves for End-Plate Connections," *Journal of the Structural Division,* ASCE, Vol. 105, ST1, January 1979.

18.26 M. J. Frye and G. A. Morris, "Analysis of Flexibly Connected Steel Frames," *Canadian Journal of Civil Engineering,* Vol. 2, No. 3, September 1975.

18.27 S. W. Jones, P. A. Kirby, and D. A. Nethercot, "Effect of Semi-Rigid Connections on Steel Column Strength," *Journal of Constructional Steel Research,* Vol. 1, No. 1, September 1980.

18.28 C. W. Lewitt, E. Chesson, Jr., and W. H. Munse, *Restraint Characteristics of Flexible Riveted and Bolted Beam-to-Column Connections,* Engineering Experiment Station Bulletin 500, University of Illinois, Urbana, 1969.

18.29 J. E. Regec, J. S. Huang, and W. F. Chen, *Test of a Fully-Welded Beam-to-Column Connection,* Fritz Engineering Laboratory Report No. 333.21, Lehigh University, Bethlehem, Pennsylvania, September 1972.

18.30 J. D. Griffiths and J. M. Wooten, Discussion to "A Fresh Look at Bolted End-Plate Behavior and Design," by N. Krishnamurthy, *ASIC Engineering Journal,* Vol. 16, No. 2, Second Quarter 1979.

18.31 American Institute of Steel Construction, *Steel Construction Manual,* 8th ed., Chicago, 1980.

18.32 H. Agerskov, Discussion to "A Fresh Look at Bolted End-Plate Behavior and Design," by N. Krishnamurthy, *ASIC Engineering Journal,* Vol. 16, No. 2, Second Quarter 1979.

18.33 W. McGuire, Discussion to "A Fresh Look at Bolted End-Plate Behavior and Design," by N. Krishnamurthy, *ASIC Engineering Journal,* Vol. 16, No. 2, Second Quarter 1979.

18.34 N. W. Rimmer, Discussion to "A Fresh Look at Bolted End-Plate Behavior and Design," by N. Krishnamurthy, *ASIC Engineering Journal,* Vol. 16, No. 2, Second Quarter 1979.

18.35 J. D. Graham, A. N. Sherbourne, A. N. Khabbaz, and C. D. Jensen, *Welded Interior Beam-to-Column Connections,* AISC, New York, 1959.

18.36 W. F. Chen and I. J. Oppenheim, *Web Buckling Strength of Beam-to-Column Connections*, Fritz Engineering Laboratory Report No. 333.10, Lehigh University, Bethlehem, Pennsylvania, September 1970.

18.37 W. F. Chen and D. E. Newlin, "Column Web Strength in Steel Beam-to-Column Connections," *Journal of the Structural Division*, ASCE, Vol. 99, ST9, September 1973.

18.38 J.-J. R. Cheng and J. A. Yura, "Local Web Buckling of Coped Beams," Journal of Structural Engineering, ASCE, Vol. 110, No. 10, October 1986.

Author Index

Subject Index